Engineering for a Finite Planet

Peter Davey

Engineering for a Finite Planet

Sustainable Solutions by Buro Happold

Birkhäuser
Basel • Boston • Berlin

Layout and cover design:
Peter Willberg and Rory McCartney, London

Library of Congress Control Number: 2007933207

Bibliographic information published by the
German National Library
The German National Library lists this publication
in the Deutsche Nationalbibliografie;
detailed bibliographic data is available
on the Internet at http://dnb.ddb.de

2009 Birkhäuser Verlag AG
Basel • Boston • Berlin
P.O.Box 133, CH-4010 Basel, Switzerland
Part of Springer Science+Business Media

Printed on acid-free paper produced from
chlorine-free pulp. TCF ∞
Printed in Germany

ISBN 978-3-7643-7220-0

www.birkhauser.ch

9 8 7 6 5 4 3 2 1

CONTENTS

INTRODUCTION

I have often thought that buildings should be more like trees. Not that they should look like trees, but that they should behave like trees, with integrated structural and services organisations, and systems of shading and ventilation that alter according to the seasons. Obviously, the metaphor can be forced too far. For instance, buildings do not draw in or emit gases in the same way as trees; nor do they grow throughout their lives. But the tissues that form the structure of the tree are also the conduits for water and dissolved minerals from the roots to the leaves, and (some of them) act as conductors of sugars generated by photosynthesis in the leaves back to other parts of the tree.

Living organisms are quite the most efficient systems we know; in them, there can be no disjunction between the elements necessary for life. They are efficient because they are integrated, and are honed by evolution to be as economical as possible with the materials from which they are made and with substances like air, water and nutrients needed for life. Like arboreal tissues, their components usually perform two or more functions simultaneously. It is this kind of economical efficiency that buildings should aim for in an age when global warming, caused by uneconomical (in biological terms) human actions, is becoming more and more clearly the prime threat facing humanity. We need organic architectures in the true, not stylistic, sense.

Buro Happold was one of the first firms to set out to blur the boundaries between the traditional disciplines of building design. To achieve more integrated solutions than those that emerge from the traditionally divided responsibilities of the disciplines, some compromises have to be made. For instance, structural engineers since the Middle Ages have traditionally striven to achieve the lightest and most slender result compatible with safety. But in some integrated designs, a heavy structure may be necessary to provide thermal inertia or acoustic isolation: the principle must be to optimise the whole, rather than any particular aspect.

To generate work of this kind, there must be creative relationships between members of the design team. Such matters are always difficult to analyse, not least because relationships vary from job to job. Architects are often reluctant to reveal to what extent they are indebted to their consultants. A further reason for obscurity is the difficulty of disentangling exactly what goes on in the process of design, which is an iterative, evolutionary and syncretic process with inputs from several sources. Buro Happold's founding partner Ted Happold believed that 'the best work is done by the most diverse group of talents who can still live together'. In most cases, the nature of the design team and its processes can only be guessed at by inspection of the completed work.

As this book tries to show, Buro Happold plays an important part in the genesis of most of the buildings with which it is involved. From the first, when Ted started to work with Rolf Gutbrod and Frei Otto in Saudi Arabia, the firm had always been committed to explorative design. The range is impressive, varying from skyscrapers like the Al Faisaliah Centre in Riyadh (with Foster + Partners) to major enclosures like the Millennium Dome (with Richard Rogers – however disastrous this was as a political gesture, it was a brilliant engineering and aesthetic success).

Before the advent of fast, large-capacity computers, shell and tensile structures were impossible to calculate in detail, and the firm was involved with experiments using bubbles and fabric models to predict how the new geometries would work structurally. Buro Happold has never been afraid of original research, and in this respect they are very different from most engineers – and from architects who often say that they are 'doing research', by which they mean nothing more than messing about with shapes. Buro Happold's research on the other hand is truly scientific in the Popperian sense – its findings can be tested and, if necessary, refuted.

Original research continued (and continues) on many fronts. For instance, at Hooke Park College in rural Dorset, working with Richard Burton, Ted Cullinan and John Makepeace, Buro Happold embarked on exploration of the potential of green softwood forest thinnings allied with new forms of epoxy joint to demonstrate innovative ways of building that could revolutionise the forestry of the northern hemisphere. In Wales, at the Wales Institute for Sustainable Education (WISE), the relevance of ancient techniques of earth construction to today is being practically investigated.

Early in the firm's existence, it became clear to Ted and his colleagues that, by offering services that spanned the traditional gap between structural and services engineering, more integrated and economical design could be achieved. Perhaps the harsh climate of Saudi Arabia was initially responsible for pressure-cooking the mix, but the thinking quickly evolved to deal with temperate climates, for instance in the pioneering green office towers at Essen with Christoph Ingenhoven, and again with Ingenhoven for the Commerzbank Headquarters competition at Frankfurt where their entry was placed second to that by Foster + Partners. In these, ventilating façades allow individual users to adjust their own office temperature and fresh air intake; sophisticated building management systems control energy consumption of the whole building complex. The concrete structures of the towers act as thermal flywheels for balancing internal temperature and reducing dependence on fossil fuels. The approach has been developed with much sophistication in the Genzyme Headquarters in Cambridge, Massachusetts (with Behnisch, Behnisch & Partner). There, structure, heating and ventilation act with weather conditions (moderated by external walls that alter according to incident radiation) to generate an interior full of fresh air and light – often sunlight.

Such thinking now permeates the work of the practice, but it is not confined to high-tech, highly glazed buildings. The merits of high thermal inertia and heat capacity in modified traditional construction, and of scientifically re-evaluated age-old considerations like orientation and cross-ventilation, have been shown to offer multiple possibilities for creation of energy-economical buildings, for instance the offices for the Building Research Establishment (with Feilden Clegg Bradley Architects) and the Wessex Water Operations Centre (with Bennetts Associates).

Working with masonry has not been limited to construction of energy-efficient buildings. One of the most dramatic recent projects has been the British Museum Great Court (with Foster + Partners). The original 19th century museum by Robert Smirke was designed as a rectangular block with a square courtyard in the middle; some years later, Robert's brother Sydney inserted the famous round Reading Room into the court and surrounded it with enclosed book stacks. The architects decided to remove the stacks and recreate the court under a glass roof. But the Smirkes were not particularly concerned with accuracy within the court, and the new glazed structure had to allow for numerous asymmetries, so that each component in both steel and glass had to be differently sized – a technique possible at large scale only since the advent of high-powered calculating.

Advances in computing have also allowed modelling of façades to study their probable performance under varying climate conditions and in fire, a skill that the firm has recently added to its repertoire. Similarly, new forms of calculating have enabled masterplanning studies to demonstrate inter-reaction of different elements. The planning strategy for large-scale projects typically relate hydrology, ecology, waste management, transport, utilities and infrastructure. All these are to be organised to maximise sustainability. Planning of this kind begins to suggest further developments in architectures that can be truly called organic because they begin to learn from living systems.

If it is to survive, a firm must be organic too. Like a tree, it must constantly adapt and change if it is to be successful. The seed that was planted by Ted Happold and his colleagues some 30 years ago continues to grow with vigour, in the understanding that technology is always about change and that the art of engineering is to use technological possibilities to enhance human life. As Ted said, engineering 'is intensely creative; at its best it is art in that it extends people's vision of what is possible and gives them new insights'. Long may Buro Happold continue to pursue the art with the generous creative imagination of its founders.

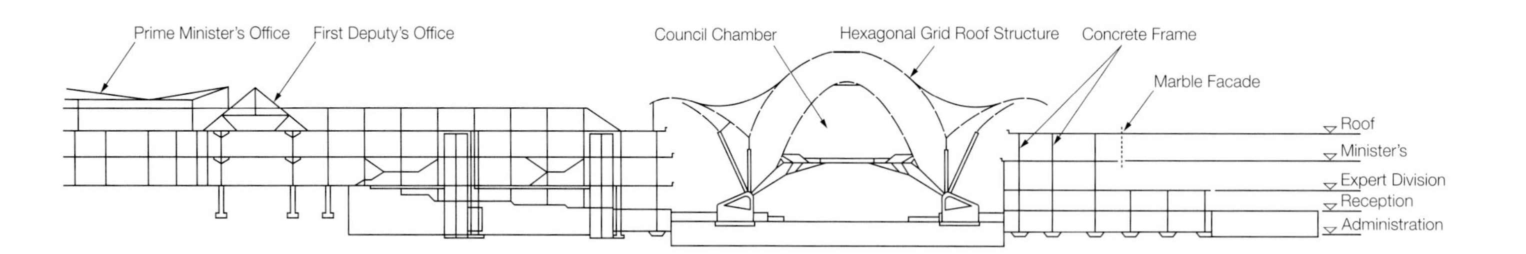

Section of the Council of Ministers building

KING'S OFFICE AND COUNCIL OF MINISTERS

Location: Riyadh, Saudi Arabia
Date: 1974
Architect: Rolf Gutbrod with Frei Otto
Client: Kingdom of Saudi Arabia
Buro Happold services: Joint civil and structural engineering with Ove Arup & Partners

Design for a government campus on 1 square kilometre adjacent to Riyadh's Diplomatic Quarter. The project was aborted after the death of King Khaled. Two factors of major importance resulted, however: it acted as a catalyst and key factor in the early growth of Buro Happold, and provided support for valuable engineering research into the highly complex roof structure, shade structures and translucent marble façades required.

Image on previous page: View inside the Tower of Steel at the Royal Armouries Museum

Significantly, Buro Happold began in Saudi Arabia. In 1974, Ove Arup & Partners were appointed with architect Rolf Gutbrod to be designers for Kocommas (the King's Office, Council of Ministers and Majlis al Shura (Consultative Council) Central Government complex in Riyadh). When Ted Happold left Arup to set up Buro Happold in 1976, Gutbrod asked the new firm to work jointly with Arup on structural and civil engineering aspects of the project. Schmidt Reuter was services consultant and architect Frei Otto was commissioned for lightweight structures. Perhaps because professional boundaries were much more fluid in Saudi Arabia than in a developed country, relationships between the firms that made up the design joint venture team were more flexible than usual.

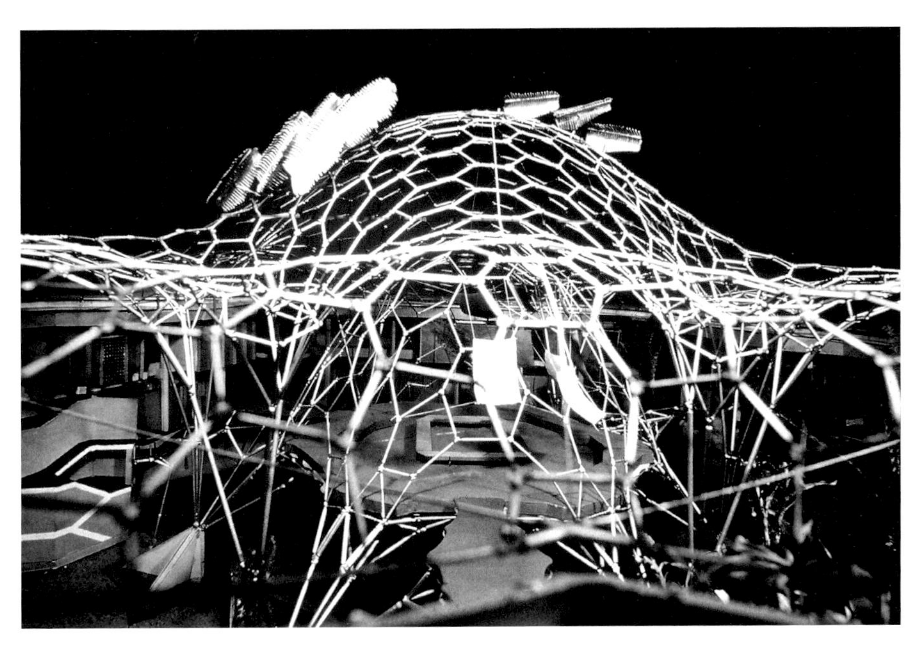

Frei Otto's model of the Council shell roofs, showing shading devices

Erectors climb Sports
Centre cable net

Day of completion
of cable net

The extreme climate helped foster the creative mix. For instance, the team evolved a form of translucent marble cladding to moderate intense desert heat and light which would have been hard to achieve across conventional professional disciplines at the time. Unfortunately, the death of King Khaled led to a new Saudi government policy and the project was abandoned, though the marble cladding was experimentally mocked up and tested, and much of the aluminium and timber umbrella structure designed to shade car parks was built and used in the development that replaced the original design. Boundaries between conventional structural and services engineering had been creatively blurred, and what seemed to be the seeds of new forms of building for the area were being sown. (Sadly, buildings in the Middle East have in general taken a different course, and the area, so rich in ambient energy, scandalously wastes vast quantities of fossil fuel to cool glass curtain-walled developments by using crude forms of air conditioning.)

Outer membrane of Sports Centre now in position

KING ABDUL AZIZ UNIVERSITY SPORTS CENTRE

Location: Jeddah, Saudi Arabia
Date: 1979
Architect: Rolf Gutbrod with Frei Otto
Client: KAA University
Buro Happold services: Structural engineering

The 90 x 120 metre Sports Centre provides a clear indoor playing area for up to three tennis courts, a range of basket ball and gymnastic requirements and a spectator seating tribune for up to 2000 spectators with VIP and changing facilities beneath the stands. This facility is housed within a doubly-curved double membrane structure supported by a prestressed galvanised steel cable net of 12 millimetre wire ropes and four 800 and 600 millimetre diameter tubular masts along either long side. The cable net is anchored to the ground by a concrete ring beam on ground anchors.

The space beneath this translucent double membrane is air conditioned but the need for cooling is reduced by the intermediate space between the two membranes. The cable net and membranes were prefabricated to accurate precalculated patterns in Switzerland and shipped to Jeddah for rapid assembly and erection on site.

The continuing informal relationship between Buro Happold and Frei Otto (who specialised in lightweight structures) evolved from Kocommas. One of its earliest results was the King Abdul Aziz University Sports Centre in Jeddah, completed in 1979. A 27-metre-high, 120 × 90 metre tent-like structure was evolved to enclose 9,500 square metres. The tent fabric had to modify the intense light and desert winds and storms. Using soap bubbles, then accurate fabric models, the designers evolved a cable net structure that could support the fabric and resist imposed loads. (At the time, desktop computers lacked enough storage capacity and were too slow to solve the non-linear equations necessary to analyse the complex shapes of fabric structures.) In fact, the membrane is double, and the space between the two layers acts as a climate buffer, incorporating external air taken from low level, and mixing it with treated air before it is expelled at the top of the structure. Again, traditional boundaries between structural and environmental engineering are dissolved in creative interaction.

Aerial view of inhabited wall and internal and external rose structures of Tuwaiq Palace

The centrepiece of the palace is the beautiful Heart Tent, a stainless-steel cable structure covered with hand-painted glass by artist Bettina Otto.

Another Saudi Arabian building, the Tuwaiq Palace in Riyadh, designed with architects Omrania and Frei Otto, further developed the environmental potential of fabric forms, this time in conjunction with heavy structure. A thick concrete wall (actually a long, thin building) faced with local stone curves to make a courtyard, protecting the space it defines from wind and providing an anchorage for translucent fabric structures that cover the larger enclosed spaces. The tents devised by the Happold/Otto collaboration are carried on grid net cable structures similar in many ways to the ones used in the Jeddah sports hall. In principle, the wall contains service spaces, while the tents are the public areas. The heavy wall provides a thermal flywheel that helps modify the temperatures within the tents, reducing cooling loads and power consumption.

View of external translucent rose structures against Tuwaiq Palace wall from Wadi Hanifah

TUWAIQ PALACE

Location: Riyadh, Saudi Arabia
Date: 1986
Architect: Omrania and Frei Otto
Client: Arriyadh Development Authority
Buro Happold services: Building services, structural engineering, quantity surveying, project management, site infrastructure

Within the Diplomatic Quarter in the north-west of Riyadh is Tuwaiq Palace, a complex originally designed as a diplomatic club but now used by the Saudi government as a hospitality centre. The demands of the brief were for a sinuous stone-walled design to create a stable protection from the harsh environment, while the tented structures rising from it respect local culture and tradition.

Translucent PTFE glass fibre membrane cables anchored back to the wall structure

In temperate England, the shading properties of fabrics were explored in the 1986 Velmead Infants School in Fleet designed with Hopkins Architects for the Hampshire County Council (by far the most architecturally creative public authority in Britain of the period). The site is protected to the north by a belt of evergreen conifers, while to the south are splendid views of heathland punctuated by clusters of deciduous trees. A linear plan was adopted, arranged parallel to the trees. In the middle of the plan is a central top-lit internal street; full-height glass walls on the long north and south sides bring in more daylight and allow the young pupils in the offices and classrooms to look straight into the countryside. In the northern strip of the plan, offices and general spaces look toward the trees, while all the classrooms are on the south side.

Children playing in front of structure to main entrance

Detail of connection nodes to steel tubular columns

VELMEAD INFANTS SCHOOL

Location: Fleet, Hampshire, UK
Date: 1986
Architect: Hopkins Architects
Client: Hampshire County Council
Buro Happold services: Structural engineering, building services

Provision of an open-plan, metal and glass modern building for an infants school in Fleet, close to Farnborough. The site itself is on heathland and the building and play areas were designed to make minimal impact on this beautiful countryside. Full-height windows allow as much natural daylight as possible into the classroom and also give users a sense of space extending the learning environment beyond the bounds of the building.

Solar heat gain in the classrooms was clearly going to be a problem, and stressed PVC fabric sails extend from the steel structure at eaves line. The arrangement shades the glass from hot, high-altitude summer sun, while allowing winter sunshine to penetrate the classrooms. To protect children in the winter from cold radiation from the glass, a 2-metre strip inside the glazed wall is devoted to wet activities and storage of outer clothes. Except in the lavatories and kitchen, natural ventilation is used throughout, with fresh air being drawn into the building through glass louvres in the long glazed walls and stale air being expelled by convection through high-level vents over the central spine. An underfloor hot water system heats the whole building and provides appropriate thermal conditions for small children, who spend much of their time on or near the floor. Velmead has performed well in practice and its visual and intellectual clarity is a striking demonstration of creative interaction between engineer and architect. This is something that Ted Happold and his partners strove to foster throughout their working lives and most obviously since Ted held the chair of Building Engineering at the University of Bath.

Canopies shade external classroom areas and full-height glazing

Both in practice and at the university, Ted was keen to promote research into new ways of building. One of the most radical was the use of softwood thinnings: these are the young trees that are removed from a forest so that the others can grow to strong maturity. Normally, they are used for firewood or perhaps fence posts. But John Makepeace, then the chairman of the Parnham Trust, which he set up to explore uses of timber, wanted to discover the potential of green roundwood thinnings in building. Makepeace commissioned Buro Happold to begin experiments at Hooke Park in Dorset near Chichester with Frei Otto and Richard Burton of architects Ahrends, Burton and Koralek.

The thinnings were used in two ways: as flexible poles in roofs and as spars in tension, bending or compression, forming rigid frames, floors and platforms. To make the system work, new joints had to be evolved, particularly ones that could cope with tension. The Buro Happold solution worked out with Bryan Harris, Professor of Materials at the University of Bath was to use a threaded bar embedded in a two-part epoxy resin in stepped holes in the ends of poles. Added to the two-part resin is cellulose microfibre filler that improves the compound's properties and increases the elasticity of the joint. A similar compression joint was also evolved to offer localised strength against crushing.

End windows overlook woodland from Hooke Park Workshop

WORKSHOP, HOOKE PARK COLLEGE

Location: Dorset, West Sussex, UK
Date: 1988
Architect: Richard Burton of Ahrends, Burton and Koralek with Frei Otto
Client: The Parnham Trust/Hooke Park College
Buro Happold services: Structural engineering, civil engineering, building services, geotechnical engineering

Temporary polypropylene sleeves to protect preservative in round wood arch members

Hooke Park College is a training centre in the heart of the Dorset woodland which aimed to educate people about the uses of green roundwood. The structures were designed by teams dedicated to pushing the boundaries of building with wood. The Workshop, a collaboration by Frei Otto, Ahrends, Burton and Koralek and Buro Happold, experiments with bending green wood and carrying loads across large spans on small-diameter roundwood beams.

The refectory, by the same team, is a prototype for a house, in which the structure hangs like a tent on four A-frames. The complex of the building has been acquired by The Architectural Association for continuing use for teaching and learning.

Initially, two buildings were constructed: a prototype house and a training centre. The house has an A-frame central structure from the apex of which bearing members of the roof (60–90 millimetre diameter thinnings) are draped in catinary curves. An inner membrane of canvas carries 75 millimetres of mineral wool insulation, and the reinforced PVC-coated polyester roofing membrane is covered with local gravel to keep out the elements. Rooms in the house are domestic in scale and are arranged on each side of the central corridor that runs under the A-frame, but in the training centre, large spaces were required. A 42.5-metre-long structure formed of three shells each spanning 15 metres was therefore created. Shells were created by pulling opposing pairs of poles into the required curvatures on site with ropes and then joining them at the top with prefabricated crown arch members. An inner membrane of PVC-covered woven fabric is separated from the outer membrane (which is similar to the one used for the house roof) by 75 millimetres of mineral wool. The pole structures are braced by the inner membrane, which was laid on the bias to create shell action between adjacent laths.

Hooke Park's buildings demonstrated the viability of thinnings in building, and showed the environmental advantages of using such timbers close to their source (the complex is in the middle of the wood that provided its poles). If used on a large scale, the techniques could revolutionise the softwood forestry worldwide. Buro Happold's commitment to enquiry and new ideas had extended to sourcing and manufacture of building components.

WESTMINSTER LODGE, HOOKE PARK COLLEGE

Location: Dorset, West Sussex, UK
Date: 1998
Architect: Edward Cullinan Architects
Client: The Parnham Trust/Hooke Park College
Buro Happold services: Structural engineering, civil engineering, building services, geotechnical engineering

Westminster Lodge is an eight-roomed student residence and seminar building, constructed from roundwood logs varying in thickness from 150–175 millimetres and set on concrete pad foundations. The roof comprises 90–110 millimetre diameter roundwood poles, scarf-jointed into continuous lengths to make a double-layered timber grid, which covers the 9 metre square seminar room in a shallow curve and supports the turf roof.

Exterior of Westminster Lodge

Sitting room area within Westminster Lodge student residences

While Hooke Park is an example of the use of a radical new technology, the Egg Theatre in Bath (with architects Haworth Tompkins) is an example of radical re-use of existing techniques. The extremely successful Theatre Royal wanted to open a new performance space for children, young people and families. On the corner of the Theatre Royal site, a listed Victorian school had been converted into a cinema that over the years had become shabby and unloved. The theatre acquired the place and the architects gutted the old building, leaving only its stone-faced walls.

Working with FUSE, a group of 6–18-year-olds, as well as with a more conventional client body, architects and consultants evolved a concept in which the new theatre would be a separate structure from its enclosing Victorian shell. The 120-seat auditorium is elliptical, with stalls, balcony and upper seating gallery, but it is a flexible space and, apart from being used conventionally, it can form a theatre in the round, have a flat floor or even be arranged on a transverse axis. The annular volume between auditorium and the 19th century walls is occupied by ingeniously inserted green and dressing rooms, foyers and service spaces. A thin concrete slab caps the space and relates the auditorium (itself supported by a steel structure) to the outer walls; it also provides the support structure of the acoustic floor of the topmost space, the rectangular rehearsal room under the building's new stressed-skin timber roof. The auditorium is mechanically ventilated from plenums under seating in stalls and balcony, while the rest of the building is ventilated naturally through the Victorian windows. Old technology and new act in harmony just as, visually, the Victorian building cradles its precious egg.

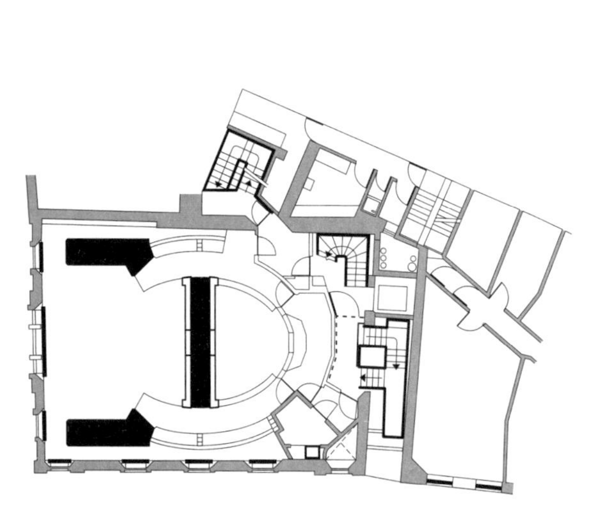

Plan of mezzanine floor with lighting and control room above Egg Theatre

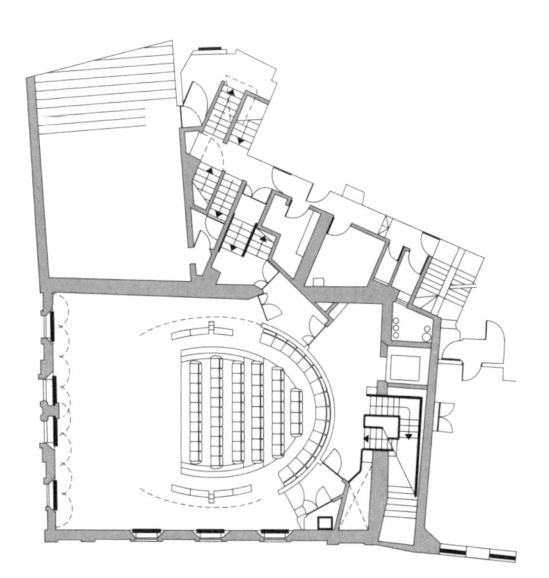

First floor plan of the Egg Theatre within the former walls of the earlier building

Old and new also work together at the Royal Armouries in Leeds. The royal and state collections of arms and armour had long been exhibited in the Tower, London's Norman castle, but they had become so enormous that a new museum was essential if more than a tiny fraction was to be shown. A site in Leeds was chosen, at Clarence Dock on the River Aire, and Derek Walker Associates were appointed as architects, with Buro Happold as structural and services consultants. The curators were convinced that the museum's internal atmosphere should resemble that of the Tower, which is protected from temperature and humidity fluctuations by very thick medieval masonry walls. Air conditioning was to be avoided wherever possible. So the Leeds museum has a massive envelope of exposed brickwork and concrete and, in most of the galleries, negligible exposure to sun and daylight. (The exception to this is the Tower of Steel, in which armour, pikes, muskets, swords and the like are glitteringly displayed within a concrete-cored glass and stainless-steel tower.)

Internal view from stage of the120-seat Egg Theatre with sound-proof windows able to let in natural light

THE EGG, THEATRE ROYAL

Location: Bath, UK
Date: 2005
Architect: Haworth Tompkins
Client: Theatre Royal, Bath
Main contractor: Emerys of Bath
Buro Happold services: Structural engineering, services engineering, quantity surveyor, planning supervisor, access consultant

The Egg Theatre project involved the conversion of a redundant cinema into a 120-seat theatre, green room, dressing rooms, workshop, control room, rehearsal studio and foyer, all linking in to the adjacent Ustinov Theatre and Theatre Royal in Bath. Buro Happold started in the ground by diverting the drainage and underpinning the existing foundations, creating as much space as possible. Reinforced concrete was used to form the basement and café structure and floor to the theatre itself. Above the concrete, the solution changed to steelwork. A lighter solution was needed that could support the audience and technical areas. The structure is exposed, allowing it to become part of the technical canvas. Nothing remains of the original building except for its external wall.

The other galleries are arranged in two stacks, with a six-storey-high internal street between them. In general, galleries maintain their temperatures by relying on the high thermal mass of the building by avoiding false ceilings and introducing low-velocity cool air at the bottom of the volumes and extracting it high up. Ceilings are high to accommodate some of the exhibits (an elephant in full war armour is one of the most spectacular elements of the show). But height has another purpose: combined with the low-level, low-velocity air supply, temperature stratification allows comfort zones to be created at the lowest level (where people are), while the upper parts of the volumes need no conditioning. Where necessary, ceiling-mounted cool radiant panels can be installed to suit special conditions (for instance in the mezzanines, where imperforate glass balustrades prevent cool air spilling from floor level down into the internal street).

Tower of Steel at the Royal Armouries Museum

ROYAL ARMOURIES MUSEUM

Location: Leeds, UK
Date: 1996
Architect: Derek Walker Associates
Buro Happold services: Structural engineering, building services, fire safety, traffic and transportation

The Royal Armouries Museum in Leeds is a new home for the national collection of arms and armour. The museum caters for over 1.25 million visitors a year, and incorporates five themed galleries, two cinemas, a library, offices and workshops, as well as external areas for the horses and shows. Buro Happold provided all the structural and building services engineering, with sustainable construction and operation a vital consideration from the start.

To improve energy efficiency, an innovative displacement air ventilation system was used to maintain a comfortable internal climate while reducing reliance on air conditioning. This exciting mixed-use regeneration scheme transformed the area into an attractive and thriving waterside community and premier tourist destination.

Such complex control of interior atmosphere necessitated full integration of structure and services design. So the team evolved a special H-section beam (an evolution of the earlier tartan service grids) that, coupled with double columns, allows service runs to be led freely through the building. Sadly, such ingenuity in detail and construction is not to be seen throughout the museum, for it was one of the first experiments in the British government's Public Private Partnership (PPP), an early version of the Private Finance Initiative (PFI) which has since produced a rash of mediocre, badly detailed, publicly funded buildings throughout the country. It is a tribute to the power and integrity of the basic design of the Armouries that its intentions are still so clear and its spaces work so well. Paradoxically, such clarity could only be achieved by blurring the boundaries of traditional professional disciplines.

Gallery space at the Royal Armouries Museum

LOW-ENERGY SYSTEMS

One of the earliest experiments in boundary blurring outside Arabia was the new office building for the Building Research Establishment (BRE) at Garston near Watford (with architect Peter Clegg of Feilden Clegg Bradley). Buro Happold did the structural engineering and Max Fordham and Partners acted as environmental consultants. Concrete upper floors have a sinusoidal ceiling profile maximising the surface area of the ceilings so that they can modify the temperature of the spaces below by using the thermal flywheel effect. Fresh air is drawn to the middle of the plan through ducts cast into the floors. Other control and energy-saving devices include louvered screens to prevent solar heat gain, arrays of photovoltaic cells on the wall and the use of coarse aggregate made of recycled concrete taken from an outmoded building on the site (recycling material like this cuts down energy used in transport as well as providing a cheap and readily available material).

External view of Kildare Country Council office building

ÁRAS CHILL DARA, KILDARE COUNTY COUNCIL HEADQUARTERS

Location: Naas, Kildare County, Ireland
Completion: 2006
Architect: Heneghan Peng Architects with Arthur Gibney & Partners
Client: Kildare County Council
Buro Happold services: Building services

Áras Chill Dara is the new civic building for Kildare County Council, situated in the town of Naas, County Kildare, Ireland. The building has three floors of office accommodation alongside the council chamber in a 10,500-square-metre building which houses 450 staff. There is a strong environmental theme as the Council wished to pursue its own carbon dioxide reduction goals and have a showcase building to promote energy efficiency more broadly.

To achieve this, solutions include a responsive façade which maintains a comfortable internal environment, low-energy lighting, natural ventilation, high-efficiency gas-fired condensing boilers which provide low-pressure hot water to radiators, and roof-mounted solar panels to generate hot water for 70 per cent of the year.

Image on previous page: Wessex Water Operations Centre, Bath

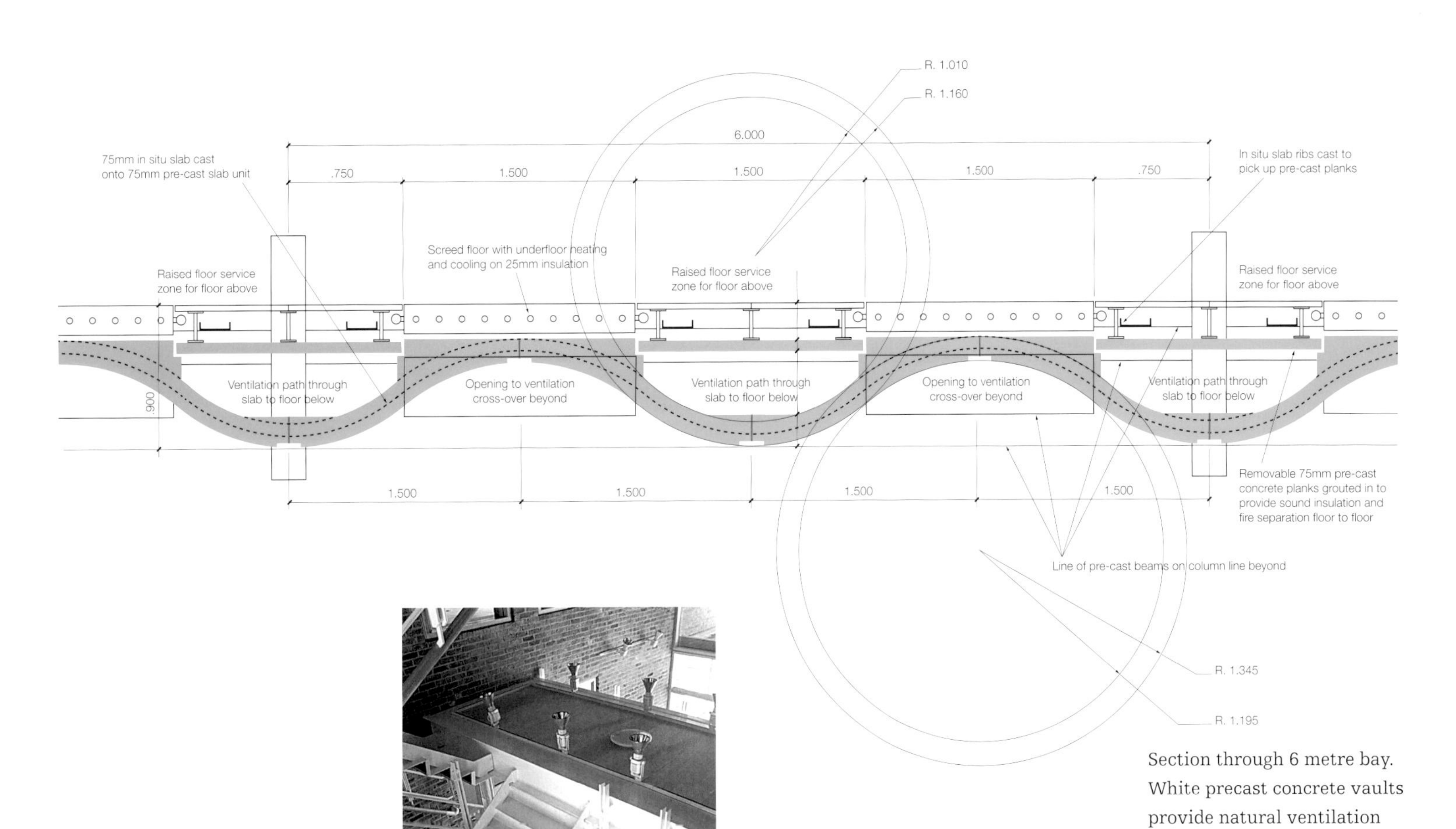

Section through 6 metre bay. White precast concrete vaults provide natural ventilation ducts and support the office floor and the floor servicing systems at BRE.

BUILDING RESEARCH ESTABLISHMENT (BRE)

Location: Garston, UK
Date: 1996
Architect: Feilden Clegg Bradley
Client: BRE
Buro Happold services: Structural engineering

This high-profile building was the first Energy Efficient Office of the Future (EOF). It combines high architectural standards with innovations in energy-efficient and environmental design, including integrating structural design by Buro Happold into the energy/environmental concept. The building has three storeys. The lower two consist of structural steelwork columns with precast and in-situ waveform floors and natural ventilation ducts built in.

This building incorporates the first use in the UK of recycled aggregates for structural concrete, used as a coarse aggregate in over 1,000 cubic metres of ready-mixed concrete for foundations, floor slabs and structural columns. Work was completed under the supervision of Buro Happold, and staff of BRE Inorganic Materials Division.

Walkways in atrium to BRE office building

BRE south elevation

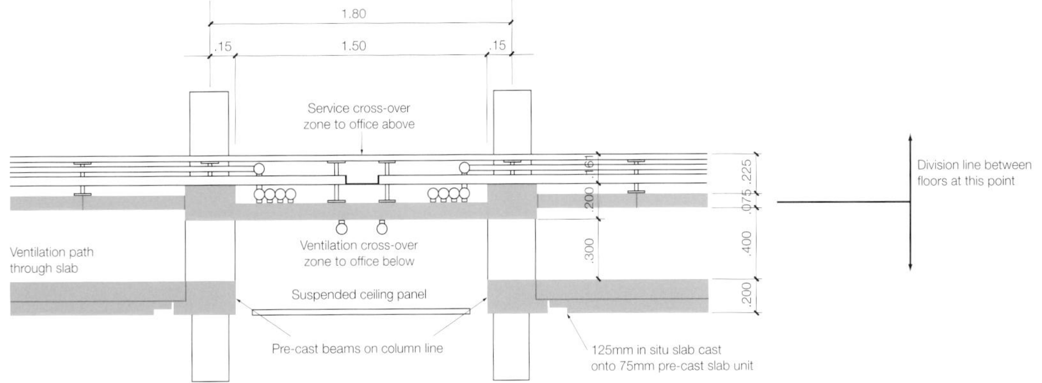

Section through corridor and vaults. Steel columns and trusses on either side of the corridor support the precast concrete vaults.

Innovation was taken further at the Berrill Environmental Building for the Open University at Milton Keynes, England (also designed with Peter Clegg in 1994, where Buro Happold provided structural, services and façades advice). The width of the building is limited to 14 metres to facilitate natural ventilation and lighting. In a complex with high IT use, heat gain is a problem, and it is combated by ventilation, external shading from solar radiation, and use of the mass of the concrete structure as a thermal flywheel. Windows can be opened manually during the building's hours of occupation and otherwise are adjusted automatically so that the mass of the concrete structure can be cooled by exposing it to night air. Electric lights are switched on and off by occupancy sensors, and are set to produce 350 lux in combination with natural light, so they are automatically dimmed or brightened according to the intensity of daylight. Sadly, the site was cramped and badly orientated, and such ideas could not be explored as extensively as they deserved.

Shading louvers to east elevation of Berrill Environmental Building

Reception and cafeteria area behind south-facing façade

BERRILL ENVIRONMENTAL BUILDING, THE OPEN UNIVERSITY

Location: Milton Keynes, UK
Date: 1996
Architect: Feilden Clegg Bradley
Client: The Open University
Buro Happold services: Structural engineering, building services

An early exploration into low-energy building design to reduce heat gain using a mix of natural ventilation, external shading and the mass of the concrete as a thermal flywheel. The Berrill Environmental Building generated much interest at the time of its completion, and set a benchmark for the development of larger structures that could successfully lay claim to being designed to a sustainable agenda. The project provided a new reception building with administration offices, conference and exhibition facilities and a refurbished 220-seat lecture theatre.

Experience of these buildings was used in the Wessex Water Operations Centre at Bath (completed in 2002, eight years after the BRE). Buro Happold worked as consultant for all aspects of the building with architects Bennetts Associates. All aspects of energy conservation, from site layout to the building management system, were radically considered. Aims included maximising natural ventilation and solar power and minimising water and electricity usage. Basically E-shaped on plan, the two-storey building follows the natural slope of the site with the bars of the E, which contain the office areas, running east-west parallel to the contours, and offering fine views of the Limpley Stoke valley. The plan's spine, which runs with the slope, is a glazed internal street unheated and naturally ventilated that relies on borrowed heat from the offices and from the mechanically heated and ventilated communal spaces: meeting rooms, restaurant, library and café on the east side of the plan.

WESSEX WATER OPERATIONS CENTRE

Location: Bath, UK
Date: 2000
Architect: Bennetts Associates
Client: Wessex Water
Buro Happold services: Structural engineering, site infrastructure, building services, geotechnical engineering, fire engineering, traffic and transport engineering, sustainability consultancy, post-occupancy evaluation

The Wessex Water Operations Centre was designed to be the greenest office building in the UK, consuming less than a third of the energy required to power a standard headquarters office building. The commitment to sustainability was embraced by the client and the whole team at every stage of the project, from inception and design through construction and operation.

This 10,000-square-metre building is E-shaped in plan with three parallel wings of flexible open-plan office accommodation linked by an enclosed street, which contains the reception space, meeting areas, library and café. The street also gives access to communal rooms on the western side: the restaurant, meeting rooms, training and community rooms and the control room (which is air conditioned).

Wessex Water Operations Centre south elevation and landscape

Again, the mass of concrete in the floors (and, here, in precast concrete wall panels) is used to balance the temperature of office spaces. Daily heat gains from people, machinery and sunshine are absorbed by the mass, which is cooled at night by drawing external air across it. Floor plates are only 15 metres wide to optimise daylight penetration and to facilitate ventilation (the bars of the E are orientated across the direction of the prevailing wind). Floor structure is lighter that that of BRE because thin precast concrete shell modules are supported on a light steel frame rather than concrete columns and beams; the structure uses 50 per cent less concrete than an in-situ one. A central axial spine beam, which also acts as a conduit for services, is formed of two steel channels set face to face with their webs perforated to provide free natural airflow across the floor plate from and to openings at eaves height on the north and south sides of the plan. The relatively light structure of the floors allows savings in foundations and in the frame itself. Floor-to-ceiling heights are great enough to allow heat rising from equipment and people to accumulate above head height, from where it is removed by natural ventilation.

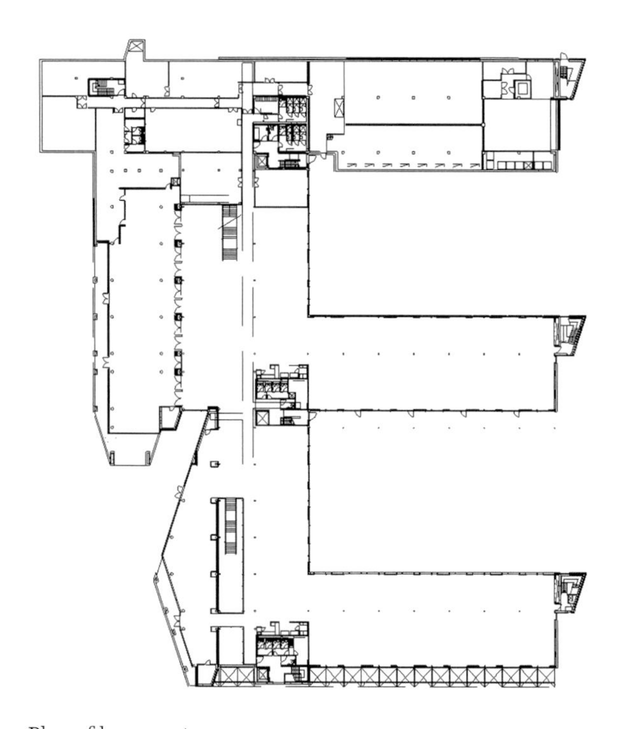

Plan of basement, ground and first floor (from top to bottom) on southward sloping site for Wessex Water Operations Centre

Northlights and south-facing photovoltaic panels above street area with cooling plant for control centre and meeting rooms

The concrete ceiling plays an important part in overall lighting strategy. Engineers and architects worked with Zumtobel Staff Lighting to evolve a luminaire that provides light to work places both directly and indirectly reflected from the concrete shells. Output from the luminaires (which use low-energy lamps) is regulated by daylight sensors within the devices themselves – sensors can be overridden by hand-held infrared devices. As part of the overall strategy that every element should do more than one job wherever possible, luminaires also house the fire alarms and have equipment for sound masking if that is required in future.

As at the Garston building, the mix for the in-situ parts of the building includes coarse aggregate of crushed concrete (40 per cent of it is from redundant railway sleepers). A terne-coated stainless-steel roof covering was chosen because of its predicted 60-year lifecycle. Other energy-saving measures include solar water heating panels, and grey and rainwater collection and recycling. The electronic building management system monitors and optimises performance of all environmental and lighting elements. User interaction is encouraged, giving people control (within limits) over the conditions of their immediate environments. The design won the highest possible rating (Excellent) for a commercial office building under the BREEAM (Building Research Establishment Environmental Assessment Method) system, and it is probably still the greenest office in Britain.

Interior view of street atria at Wessex Water Operations Centre

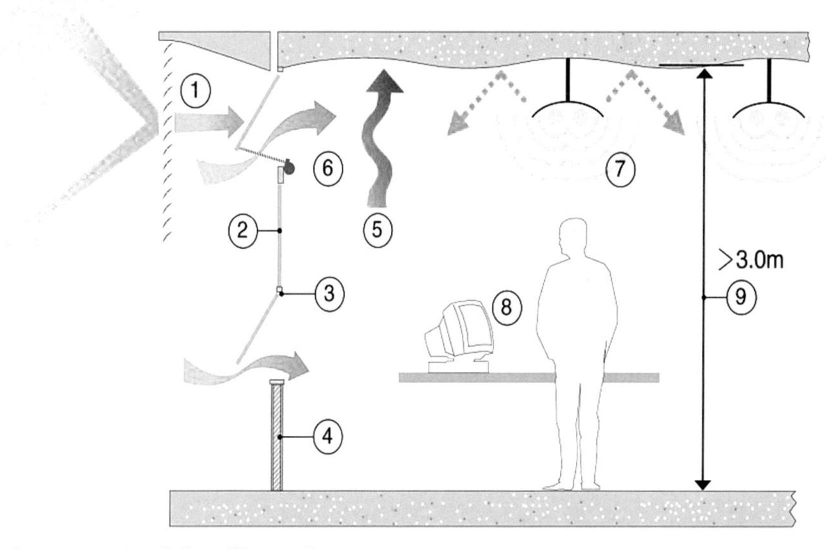

Early concept sketch for office environment

1 Rejects high-level solar gain but reflects light onto ceiling
2 High-performance light sealing and full-height double glazing
4 Manually operated low-level window
5 Exposed thermal mass
6 Automated high-level ventilation window
7 Sensor-controlled lighting
8 Table lighting
9 High ceiling

Many of the lessons learned in Wessex were developed in the Genzyme Headquarters in Cambridge, Massachusetts, completed in 2003. Buro Happold worked with design architects Behnisch, Behnisch & Partner on a commercial building that is revolutionary in the USA, where old-fashioned air conditioning and mechanical ventilation are the norm. The Genzyme Corporation, a biotechnology company, stressed that the design of its headquarters should be healthy to work in and environmentally responsible. Key aims were to achieve high levels of daylight throughout the 14-storey building, to minimise energy consumption and external impacts during both construction and use and to conserve natural resources.

The site is part of a mixed-use development built on derelict industrial land in downtown Cambridge. Its heat source is the nearby power plant that provides steam which both heats the building and cools it (by means of absorption chillers). The main frame is concrete, chosen partly for its thermal flywheel properties, but the topmost two storeys have a steel structure and form a plant-containing penthouse.

External view of
Genzyme Headquarters

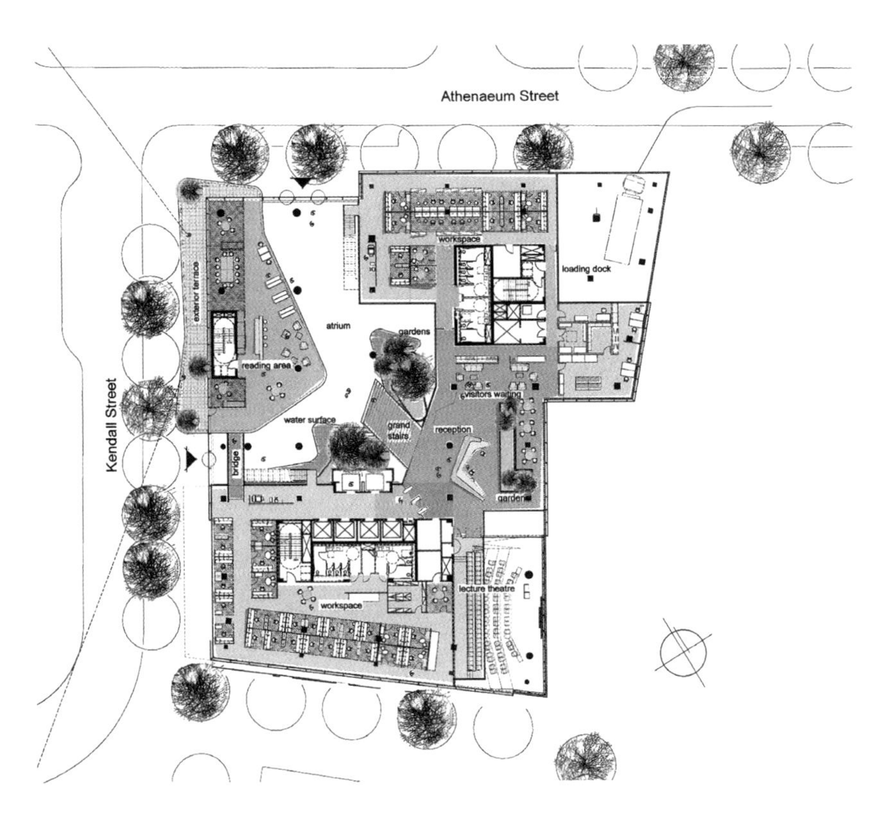

Ground floor plan

Daylight pours into the building through its glazed walls, central atrium and roof-mounted heliostat mirrors. Walls are double, with automated blinds to control solar heat gain next to the outer skin. Plants are grown in the generous (1.5 metre) interstitial space to reduce pollution and oxygenate external air. The space and its vegetation also serve to reduce noise from the surrounding streets, and the plants enhance views of the city from inside the building. Air introduced through the double wall complements the main ventilation system that draws fresh air in at roof level and distributes it tempered (heated in winter and cooled in summer) through fan coil units in the office ceiling voids. Internal carbon dioxide levels are metered and compared to those in the external atmosphere, and fresh air intake is adjusted accordingly. Return air moves across the floor plates into the 12-storey atrium, from where it rises by convection to be discharged at roof level via energy recovery devices that transfer a large proportion of its heating and cooling energy to incoming fresh air. Temperature and ventilation of individual spaces (or groups of spaces) can be varied by their occupants – for instance, perimeter offices with their windows open can have different conditions from those of spaces that look out into the atrium.

GENZYME HEADQUARTERS

Location: Cambridge, Massachusetts, USA
Date: 2003
Architect: Behnisch, Behnisch & Partner
Client: Lyme Properties/Genzyme
Buro Happold services: Structural engineering, building services, LEED assessment

Buro Happold's engineers set out to develop the building from inside out – the result is a sustainable and people-friendly office in which occupants have a high degree of control over their personal working environments. Design features include a double façade and a high atrium that creates an open spatial atmosphere while bringing natural light deep into the building.

The façade has openable windows that allow for natural ventilation and save on air conditioning costs. The ability to capture solar gains and reduce heat loss adds to comfort levels and dramatically reduces energy use. Other green features include the use of steam from a nearby power plant for central heating and cooling, rainwater collection tanks, low-flow toilets and photovoltaic solar panels to produce electricity. There are also heliostats on the roof which lights the atrium through suspended glass reflectors.

View inside the Genzyme Headquarters

This magical volume is the physical heart of the building. It is not a simple space, but a void that changes in plan as it goes up, providing a variety of terraces, balconies and bridges for communication and informal meetings. Heliostats on the roof track the sun and direct its rays through mirrors and prisms down into the atrium, where they are re-reflected into the surrounding office spaces from the lightwall, i.e vertical reflective blinds on the south side of the big volume that are automatically adjusted throughout the day to distribute sunshine throughout the interior. Artificial illumination in the office spaces, provided by luminaires in the ceilings (and sometimes by uplighters), is balanced against daylight to achieve constant working illumination of between 430–540 lux. Motion sensors in the cellular offices switch off lights if they detect no movement for 15 minutes; similar arrangements apply to zones of the public areas. All the control and emergency systems are handled and related by a comprehensive electronic building management system optimised for energy conservation, using daylight and the external climate to a maximum. When necessary, the system can be manually overridden from a conventional PC workstation. The qualities of the building were so obviously outstanding that it was awarded a very rare LEED Platinum rating.

Genzyme Headquarters
façade

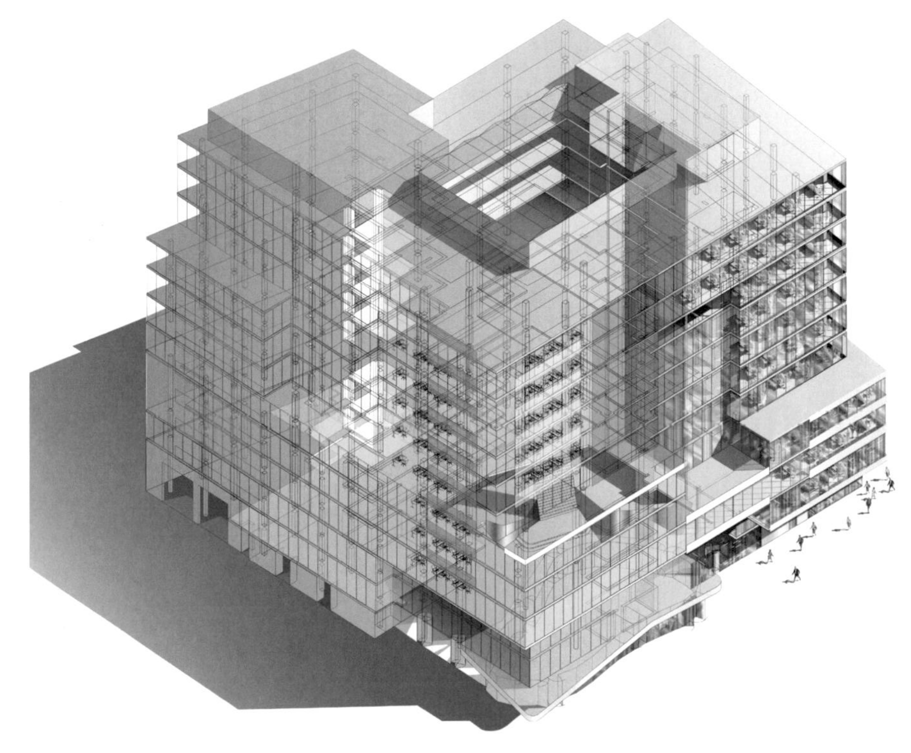

Cut-away rendering of
entrance and atrium
volume

So far, Genzyme is the most elaborate and sophisticated integrated climate control system Buro Happold has designed, but other buildings have explored a range of control systems to deal with different conditions. The Al Faisaliah Centre in Riyadh, a mixed-use development designed with Foster + Partners, clearly has to deal with a completely different climate from that of semi-temperate Boston, but an equally elaborate automated building control system was required. It had to deal not only with the extreme climate, but with the various uses of different levels of the tower. At the top is a golden glass globe housing a three-storey restaurant, and below that is a public observation deck that tops the office accommodation. Before this 30-storey project, local rules limited building heights to 15 occupied floors, so there were no precedents, and both structural and services issues had to be rethought. Partly this challenge was met by the invention of the external reinforced concrete mega-structure with its four columns clad in light metal, and intermediate K-brace transfer structures that house service plant rooms and provide walkways in the sky. This structure then supports the three zones of office accommodation that in all comprise 30 occupied floors. By concentrating office development into the tower, a generous public piazza, 80 metres long and 60 metres wide, could be created (a most uncommon urban space in Saudi Arabia). Below that is a spectacular banqueting hall under a clear-span concrete arch roof with no internal columns.

Internal view of Genzyme Headquarters atrium showing filigree concrete floor plates, balconies surrounding atrium and suspended glass reflectors

The tower façade was designed to provide maximum transparency from inside and outside. Extensive aeroplane wing-like external stainless-steel shading minimises solar gain and glare while permitting dramatic views of city and desert. Effects of insolation are further reduced by low-E double-glazed cladding units tinted to increase solar reflectance. Return air movement in the cavity combined with internal blinds also helps to reduce solar heat gain. At the base of the tower, silver-anodised aluminium panels and blade awnings reflect sunshine.

Notwithstanding all such preventive measures, a heavy cooling load is generated; it is controlled with an ice storage system. A CWC-free refrigeration plant produces 600 tonnes of ice a night, when electricity can be bought at a cheaper rate than during the day. This store of cooling energy tempers the internal climate at hours of peak demand between one and five in the afternoon. The ice stores are automatically monitored as their coolth is used, and the building management system carefully controls ice melting and refreezing to optimise energy use. As a result, electricity use is much reduced compared to a conventionally cooled building of similar size.

Glazing erection in progress

AL FAISALIAH CENTRE

Location: Riyadh, Saudi Arabia
Date: 2000
Architect: Foster + Partners
Client: King Faisal Foundation
Buro Happold services: Structural engineering, building services, civil engineering, quantity surveying, project management

The massive 245,000-square-metre mixed-use Al Faisaliah development in Riyadh is one of the most striking buildings in the Middle East. Its landmark office tower rises 270 metres over the city, soaring to a tapered point in one smooth giant arc. At its pinnacle, the tower narrows to a brightly-lit lantern topped with a decorative stainless-steel finial. At ground level of the office tower, a lobby with a petal roof links the hotel to the apartments and shopping mall. The tower, hotel and apartments enclose a landscaped plaza beneath which is an 81 × 63 metre banqueting hall, exhibition centre and two levels of parking.

The design makes extensive use of passive energy control methods: a combination of carefully chosen glass in the façade, various external shading systems and an air conditioning plant which uses ice in the cooling process serves to minimise the amount of energy drawn from the supply grid. A building management system (BMS) controls all engineering processes, monitoring and adjusting individual elements to enhance operating efficiency and living conditions.

A view of the city
showing the landmark
of Al Faisaliah

External view of atrium roof on one of the three blocks

UNIVERSITY OF PLYMOUTH, PORTLAND SQUARE BUILDING

Top-hung open windows provide natural ventilation

Location: Plymouth, UK
Date: 2002
Architect: Feilden Clegg Bradley
Client: University of Plymouth
Main contractor: Bovis Lend Lease Ltd; University of Plymouth
Buro Happold services: Building services, structural engineering, asset management, fire engineering, traffic and transport consultancy, ground engineering

This building comprises three elemental blocks all linked with a common circulation route or street at ground floor level. Each block comprises six storeys surrounding a full-height atrium, lecture theatres and teaching spaces at the lower two levels, while the upper four levels are cellular or open-plan office accommodation for academic departments and support staff. The building benefited from a series of interventions to optimise space efficiency, resulting in vertically-hung plant rooms and a high degree of on-floor prefabrication.

The building was the focus of a 12-month post-occupancy evaluation programme, instigated by Buro Happold, as part of its service to help bed in the M&E systems and tailor their control to suit occupant needs. The building achieved low-energy performance in its first year of operation.

A naturally lit street links the three building blocks.

BBC MEDIA VILLAGE

Location: London, UK
Date: 2004
Architect: Allies & Morrison
Client: British Broadcasting Corporation/Land Securities Trillium
Buro Happold services: Structural engineering, building services, façades engineering, fire safety design, acoustics consultancy, environmental and sustainability engineering, geotechnical engineering, civil engineering

The BBC is dedicated to sustainable development. Its Media Village is a 100,000-square-metre project to redevelop the site next to the existing BBC Television Centre in White City. The Village comprises the Broadcast Centre, the Media Centre, the Energy Centre, two perimeter buildings and supporting infrastructure. The development involved the entire breadth of Buro Happold disciplines, delivered in a co-ordinated and integrated way.

Design advice provided by Buro Happold's Sustainability Group allowed the development to achieve an excellent rating under the BREEAM environmental assessment scheme. Advice included an Environmental Impact Assessment, site consultancy and environmental management.

External view of Media Village buildings

External view
at nightfall

Internal view

Image on previous page:
External view of
Hazelwood Sensory
School

Schools offer particular environmental design problems: they have intermittent use; they contain a variety of functions and spaces with specific needs; they should in some places have strong links between interior and exterior, yet they must not get too hot or cold; and, though they have to cater for large numbers of people, they must not be institutional. The British government has an ambitious investment programme under which every secondary school in England will be renewed or rebuilt within a 10–15-year period. Within that overall strategy, the exemplar schools programme is an adventurous attempt to explore new approaches to school design in all its aspects, from pedagogical to environmental. Exemplar designs do not have a specific site but explore new ideas in school organisation, new relationships between schools and communities, innovative forms of curriculum and are intended to introduce and set standards for school environmental performance.

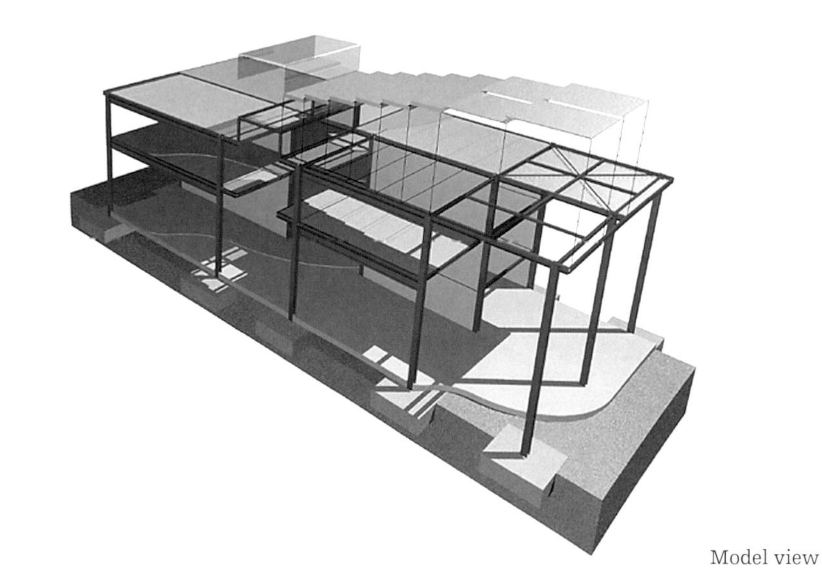

Model view

CAPITAL CITY ACADEMY

Location: Brent, London, UK
Date: 2003
Architect: Foster + Partners
Client: Department for Education and Skills; Sir Frank Lowe
Buro Happold services: Structural engineering, civil engineering, building services, fire engineering, geotechnical engineering, planning supervisor

The Capital City Academy has a national curriculum for over 1,300 pupils and it features sports and visual arts facilities recognised to be among the best in the UK. The project brief called for an aesthetically pleasing and defining building that is educationally suitable with good value for money, as well as promoting greater diversity and a radical approach to learning principles. This and the needs of all the users were taken into account when creating the engineering solutions to the building design.

The linear form of the building minimises loss of playing field area and creates links to Willesden Sports Centre. Classrooms are arranged on either side of an internal day-lit street within the building.

Interior view

A parallel programme, for city academies, is intended to create new schools in the most deprived parts of inner cities, often replacing run down and failed existing establishments. To try to generate variety, each academy has an external sponsor who contributes ten per cent of the project costs and can specify the emphasis of the school. Buro Happold has been associated with the exemplar schools and the city academy programmes from their inception in the early years of the new century.

One of the first of Buro Happold's city academies was the John Cabot City Technology College at Kingswood in Bristol, where the engineers worked with the late Richard Feilden of Feilden Clegg Bradley. A central internal street is flanked by the school's two main spaces, the hall and the gymnasium. Teaching departments are in the wings (side streets) and central facilities like the library and administration are arranged in a crescent overlooking the landscape below and stepped to form an amphitheatre. The building form was determined by the need to achieve good levels of natural lighting and ventilation. Excessive solar heat again is avoided by fixed external shading and movable blinds. Space heating and energy conservation are based on condensing boilers, underfloor warming in high spaces and very high levels of insulation in the building fabric.

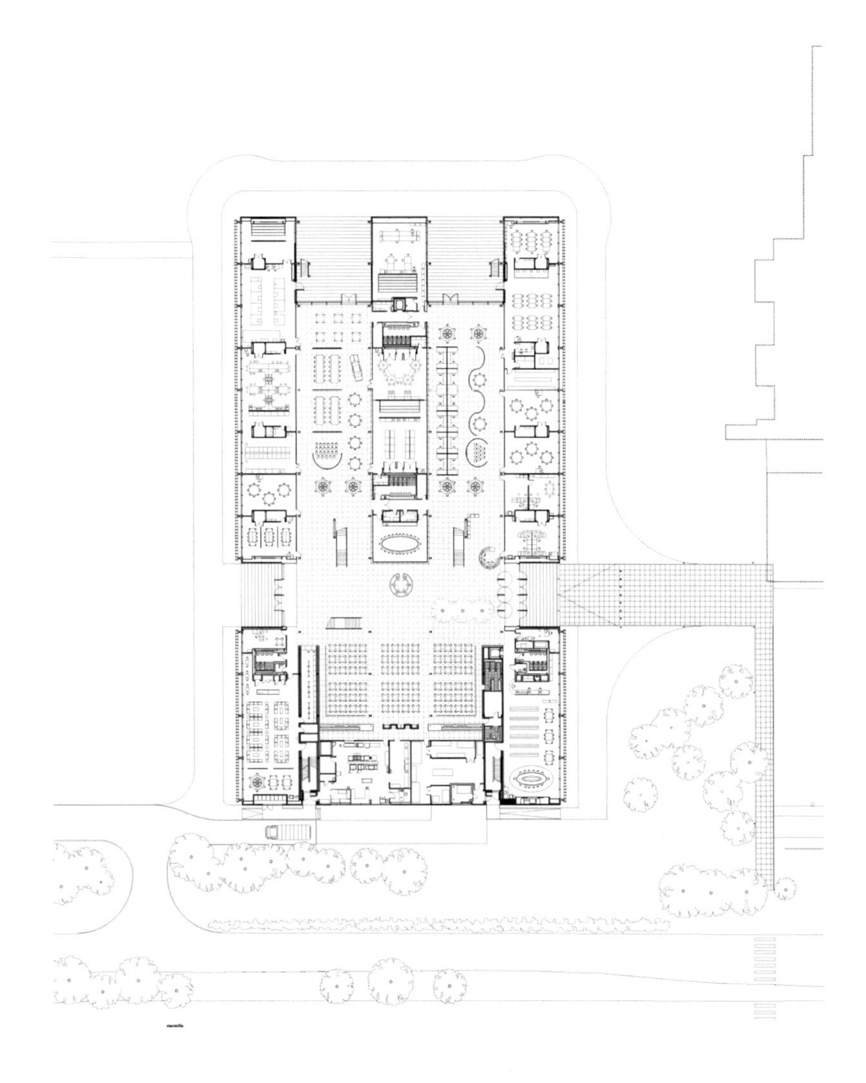

Ground floor plan

BEXLEY BUSINESS ACADEMY

Location: London, UK
Date: 2003
Architect: Foster + Partners
Client: Department for Education and Skills, 3Es Enterprises, Bexley Business Academy
Buro Happold services: Structural engineering, building services, geotechnical engineering, fire safety design

Bexley Business Academy is a three-storey educational/ business facility spanning a floor area of 12,000 square metres. The school provides education for 11–18-year-olds, with further emphasis on business skills for 1,350 students and 150 staff. An open-plan layout achieves a large degree of transparency between the students, school and community. The open-plan approach ensures that classrooms have no fixed boundaries and overlook central courtyard areas. The school was constructed in steel to provide flexibility advantages. Precast floor systems are used with exposed soffits for environmental conditioning.

Two city academies on which Buro Happold worked with Foster + Partners are at Bexley, where the emphasis is on business, and Brent, which focuses on sports, music and drama. Buro Happold found that the programmes had much in common. Both have populations of some 1,500 students, and each has to provide general teaching accommodation as well as specialised spaces for science and technology, art rooms (with kilns), music and craft rooms, national standard sports halls, IT training areas, auditoria with retractable seating, and restaurants and kitchens to provide meals at all times of the day (including breakfast). In addition, the Bexley Business Academy has a mini stock exchange. All these functions had to be linked by generous atria and circulation spaces, and they had to be created under an accelerated design and construction programme to provide results for the politicians as quickly as possible.

External view of the main entrance and vertical shading louvers

Classroom corridor overlooking central atrium space

At Bexley, an open plan ensures that general classrooms have no fixed boundaries and overlook central courts to maximise natural daylight. In Brent, with its emphasis on sport, it was essential to minimise loss of the playing fields of the Willesden High School, which the Capital City Academy replaces. To achieve a compact plan, classrooms are arranged on both sides of an internal day-lit corridor. Even though plan forms were very different, the fast programmes necessitated a form of construction that would be quick, could allow for last-minute alterations and facilitate future changes and extensions. Steel frames were chosen for their flexibility and, in both cases, precast concrete floor planks were used to provide thermal mass. In summer, the resulting slabs are exposed to cross-ventilation at night to moderate daytime temperatures. Each had a different type of precast concrete. At Bexley, precast, prestressed hollow planks made tricky last-minute alterations to the location of holes in the slab. Brent's flat-form planks had a thick two-way spanning topping that made punching holes easier, but problems in the supply chain created tolerance difficulties. In both cases, the finish of the undersides of the floors is smooth, well-finished and works both climatically and visually.

External view of Hazelwood Sensory School

Since these first academies, Buro Happold has been involved in the design of over 30 more. Of these, some have been explicitly designed as educational examples of sustainability in building. The Langley Academy of Science near Slough for instance, again designed with the Foster practice, is intended not only to reduce the environmental impact of the building, but to raise awareness of environmental sustainability in the school community. The whole building with its landscape is intended to be a huge teaching aid.

There is a biomass boiler, a ground heat pump and solar thermal collectors as sources of heat. Rainwater is harvested, grey water is treated in a reed bed and the external lighting system is powered by photovoltaics and small windmills. The biomass boiler is the leading source of space heating, with supplementary gas boilers on the roof to meet peak loads. The ground source heat pump system provides underfloor heating to the restaurant and assembly area. The solar thermal collectors provide preheat to the gas-fired water heaters. Interconnections between all these systems are made clear as part of the educational system. Similarly, the workings of the rainwater harvesting system and the water-purifying reed bed are clearly demonstrable to students.

HAZELWOOD SENSORY SCHOOL

Location: Glasgow, Scotland, UK
Date: 2006
Architect: Gordon Murray + Alan Dunlop Architects
Client: Glasgow City Council
Buro Happold services: Structural engineering, building services, disability design consultancy, ground engineering, specialist consultancy, infrastructure

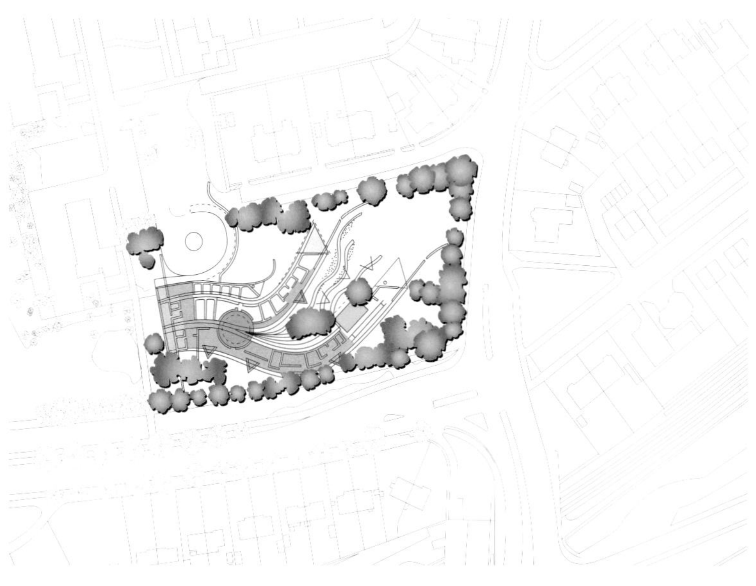

Site plan of the school

Hazelwood Sensory School in Dumbreck, a suburb of Glasgow, is a very special school for children with severe sensory impairment, the first of its kind in Europe. It has been specially designed to provide a learning environment for 52 children with additional support needs. Sustainability was key, with the building crafted from local timber and slate. The school has been designed to provide life-learning skills for children from ages four to 18, who are blind or have severe sight impairment, often combined with a hearing and mobility impairment.

New internal streets have been created with areas specifically designed to provide sensory and tactile clues to help children locate themselves within their immediate environment. The building incorporates classrooms which allow in natural light from the north and screens which open out to protected gardens to the south.

All energy controlling systems are linked by a building management system (BMS) that automatically relates internal to external conditions, introducing heat, cooling and ventilation as required by the different functions and orientations of the building. As in the Bexley and Brent academies, concrete floors form thermal flywheels, and in the classrooms external air is conducted past the undersides of the slabs by high-level windows automatically opened and shut by the BMS. Low-level opening lights are manually operated. Mechanical ventilation is largely limited to spaces like the lecture and drama theatres, the sports hall, workshops and music rooms, though there are some areas such as the restaurant that are both mechanically and naturally ventilated. The BMS relates cooling by air handling units and heating from either radiators (in the classrooms and offices) or from underfloor systems (in areas like the sports hall, restaurant assembly space and drama theatre). All areas with mechanical ventilation have heat recovery systems.

Comparative studies of several academies have demonstrated that Langley shows a reduction of 20 to 25 per cent in water use compared with a more conventionally designed school of similar size, and can demonstrate a saving of 122 tonnes of carbon dioxide – largely due to the use of biomass as the main fuel, for when it is burned the (usually) forest-sourced fuel does not add to the net amount of carbon in the atmosphere, it only returns what the trees took out in the first place.

Main entrance to St Francis of Assisi

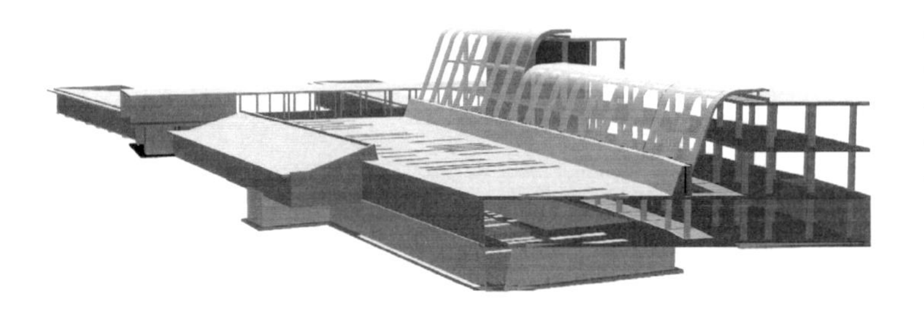

Model showing atrium with EFTE foil cushions

ACADEMY OF ST FRANCIS OF ASSISI

Location: Liverpool, UK
Date: 2005
Architect: Capita Percy Thomas
Client: Department for Education and Skills, 3Es Enterprises and the KCA Trust
Buro Happold services: Structural engineering, building services, CoSA (Computation and Simulation Analysis), geotechnical engineering, sustainability consulting

The school specialises in environmental studies, which is reflected in the sustainable design of the building. The striking architectural concept is the design approach of a turfed grass roof. The building incorporates an atrium with ETFE foil cushions on the south-facing façade to maximise the admittance of natural daylight and passive solar heating.

The building's external façades consist of thermally-massive concrete walls and exposed concrete ceiling to smooth out internal temperature fluctuation, and allow internal heat gains to be absorbed during the summer, thus reducing the need for a summer cooling system. Renewable energy solutions such as photovoltaics have been used, and a facility for rainwater recycling has been incorporated.

The Academy of St Francis of Assisi in Liverpool (designed with Capita Percy Thomas) takes the principle of the building as teaching tool a stage further. It is in one of the city's most deprived inner areas and is specifically devoted to environmental issues. For instance, all classes in the earlier years have a particular garden in which young students are introduced to environmental studies, and other subjects (for instance maths) are green-biased with projects based on the gardens. The range of devices intended to raise the level of the building's sustainability is greater than used before. Solar collectors for water heating and photovoltaic panels are used on a larger scale. A green (planted) roof adds to the thermal flywheel effect of the exposed soffits of the concrete floor slabs, and encourages the bio-diversity of the whole site. An atrium clad in inflated cushions of ETFE, a flouro-thermoplastic film with excellent durability and thermal properties, attracts solar heat to warm the building in winter.

Elevation showing
projecting window bays

View inside the
assembly hall

Paddington Academy in Westminster, London, is the first academy school to have been developed from an exemplar design (again evolved with architects Feilden Clegg Bradley). This exemplar showed how a primary school and a secondary school could be related on the same site, with shared facilities where appropriate. Its principles were also intended to be used in stand-alone secondary schools. The heart of the school is a three-storey-high covered space used for performances, dining, circulation and social activities. There are no corridors on the upper floors: upper-floor circulation is on balconies of varying widths opening into the tall volume. The two schools are linked visually through the primary hall, and by two connections at each level. Shared facilities include reception, kitchen and dining areas as well as staff and administration facilities. Secondary spaces are organised as clusters at each corner of the central space and are adaptable to many different configurations as classrooms or open-plan teaching areas.

Environmental measures behind the exemplar design include reduction of energy consumption by providing high levels of natural lighting, by ensuring that the shell is highly insulated and by introducing a very efficient ventilation system. A large measure of artificial ventilation had to be a possibility for areas with high levels of ambient noise. The system is based on an undercroft below the structure in which a labyrinth heats fresh air in winter and cools it in summer; it also ensures that the school (which is of course occupied only part-time) has a secure source of fresh air throughout the 24-hour-cycle. The ground itself acts as a thermal flywheel in addition to the concrete floors.

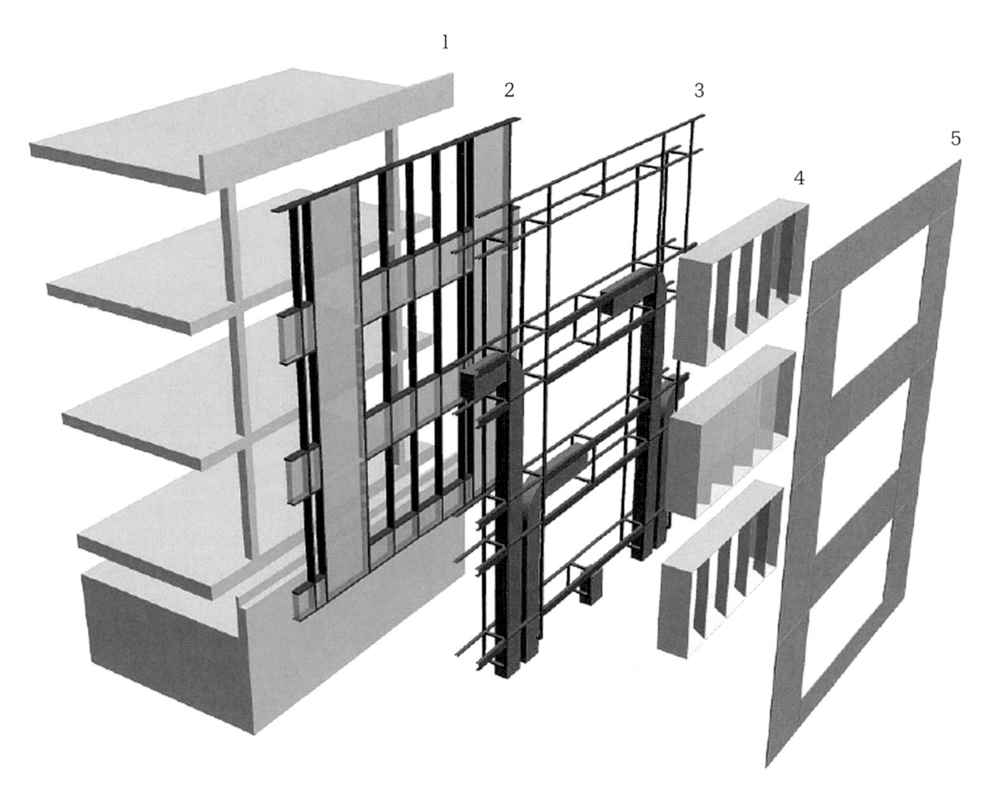

Exploded isometric of Padding Academy façade

1 Reinforced concrete structure
2 Insulated and glazed elements
3 Secondary structure containing air ducts
4 Shading louvers
5 Rainscreen to air ducts

The Paddington programme proved to be an ideal test-bed for the ideas of the exemplar proposal. It is intended to cater for 1,175 students aged between 11 and 18 (there is no primary school element in this programme). The academy, designed with Feilden Clegg Bradley, is intended to specialise in the performing arts, so it has to have at least some extremely quiet spaces, yet the site is noisy from the high level of surrounding traffic. Provision of undercroft pretreatment of fresh air has usually been regarded as expensive, but the site and hence the building footprint are constrained, making the device affordable. So there is no need for openable windows in spaces such as the theatre, multi-purpose hall, dance, drama and music classrooms, and the television and radio recording and editing suite, which all have to have very low levels of background noise. Ventilation and temperature are of course monitored by a BMS.

Low speed, highly energy-efficient fans are used to drive air through the undercroft. Vertical riser ducts at the perimeter transfer air from the labyrinth to classrooms on upper floors. The system allows much flexibility in classroom organisation. The air picks up warmth from the occupants of the peripheral spaces, and this largely heats the atrium – free. Because the system is in the undercroft, rooftop plant rooms and ducting are obviated, facilitating control of temperature and noise – and meeting the planners' stipulation that the works should not be exposed to view. As the first school to use the undercroft air handling system, Paddington's performance – and costs – will be analysed with great interest, for the principles behind the design, if they are shown to be sustainable, should be applied on a wide scale in inner city areas.

PADDINGTON ACADEMY

Location: Westminster, London, UK
Date: 2006
Architect: Feilden Clegg Bradley
Client: United Learning Trust Ltd
Buro Happold services: Building services, structural engineering, site infrastructure, security consultancy

Paddington Academy in Westminster is a new education facility to house 1,175 pupils aged 11–18 years. The school specialises in media and performing arts, and business and enterprise. The site was previously occupied by the North Westminster Community School and other community buildings. The Academy comprises two three-storey teaching blocks connected by a central atrium. The large communal spaces such as the theatre, assembly hall and sports hall connect to the central atrium linking the teaching wings. Art classrooms are located on a fourth level above the central atrium area.

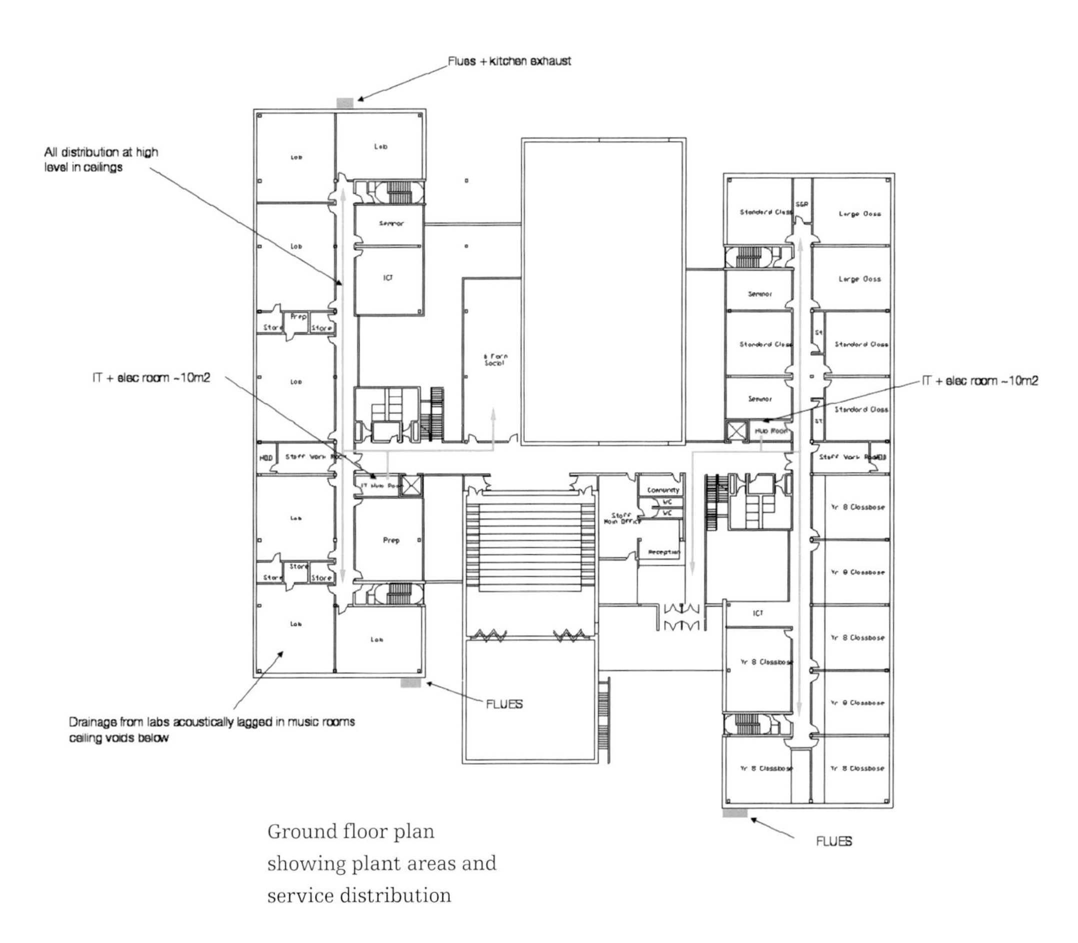

Ground floor plan showing plant areas and service distribution

A very different brief was generated by the Hazelwood Sensory School in Glasgow, where Buro Happold worked with Gordon Murray + Alan Dunlop Architects to make a special unit dedicated to providing life-learning skills for children aged between four and 18 who are blind or severely visually impaired and who often have hearing and mobility impairment as well. Areas on the internal streets have been arranged to give students tactile and sensory clues to help them orientate themselves in their immediate environments. Classrooms draw natural light from the north and open to views of the protected gardens to the south. As project architect Stacey Phillips says, the aim is to be the 'very opposite of institutional; the children can be safe and secure but at the same time have a sense of independence.'

The site is an idyllic one, parkland once occupied by a dairy surrounded by lime trees to provide shade. To avoid existing trees, the building winds across the plot from east to west, its curves set out radially and concentrically from two points in the garden area. Glue-laminated (Glulam) members form the frame of the single-storey building, and most walls, internal and external, are of prefabricated timber panels. The building is intended to be as independent of artificial heating and cooling as possible. The Glasgow climate necessitates heating in the winter months, but the only artificial cooling needed in summer will be in the IT server and telecommunications room. In summer, the trees will provide a degree of shading from solar heat gain, but the building also has shading on the south side with its windows that overlook the garden. Natural ventilation is single-sided in winter and is from trickle ventilators integrated in the fenestration.

Visualisation of external view

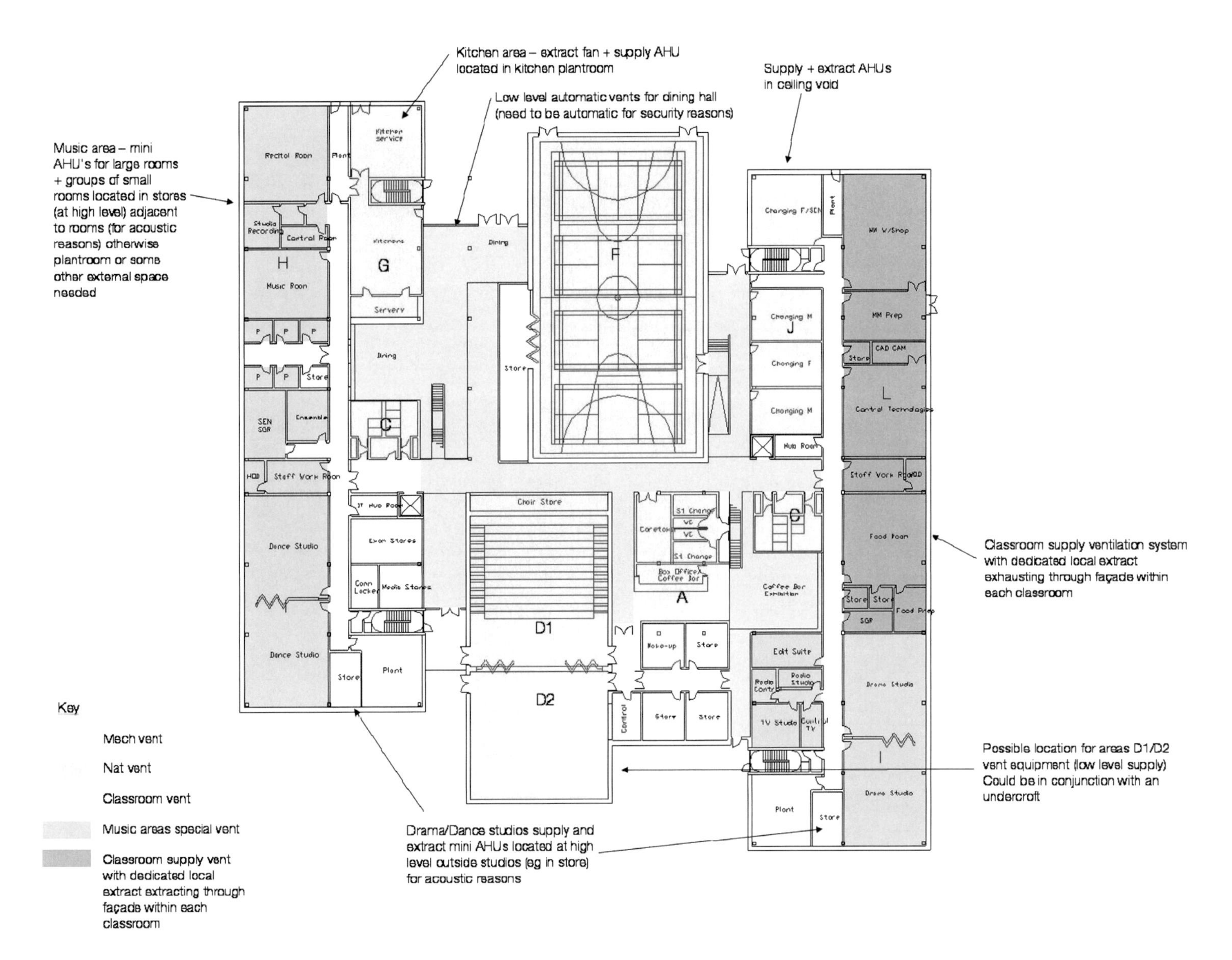

Ground floor plan showing environmental zoning

Visualisation of central atrium

LIGHTWEIGHT AND TENSILE STRUCTURES

One of Ted Happold's passions was for lightweight structures. Even before setting up his own firm, he worked with the German architect Frei Otto, the world expert in tensile structures, and such studies remain a central preoccupation of the firm today. Indeed, in addition to the Kocommas project, the Sports Complex in Jeddah and the large stainless-steel mesh aviary for the Munich Zoo with architect Jörg Gribl were among the practice's very first built projects. After the challenge to Britain's Genius Pavilion 1977 with Theo Crosby of Pentagram, one of the first of Buro Happold's British fabric structures was in London for Imagination, the visual consultants. A pair of dull brick Edwardian office blocks (one the service side of a part of the grandiose Store Street crescent) scowled at each other across a narrow alley. Architect Ron Herron, working with Gary Withers, the founder of Imagination, realised that a new internal space and new relationships could be created by roofing over the alley and connecting the two buildings by bridges.

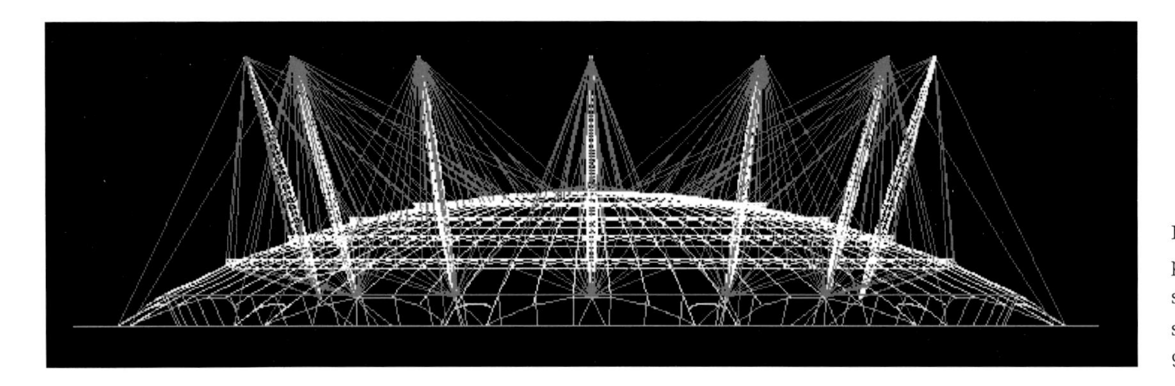

Isometric of the principal structural system to the Dome showing 12 masts, each 90 metres high

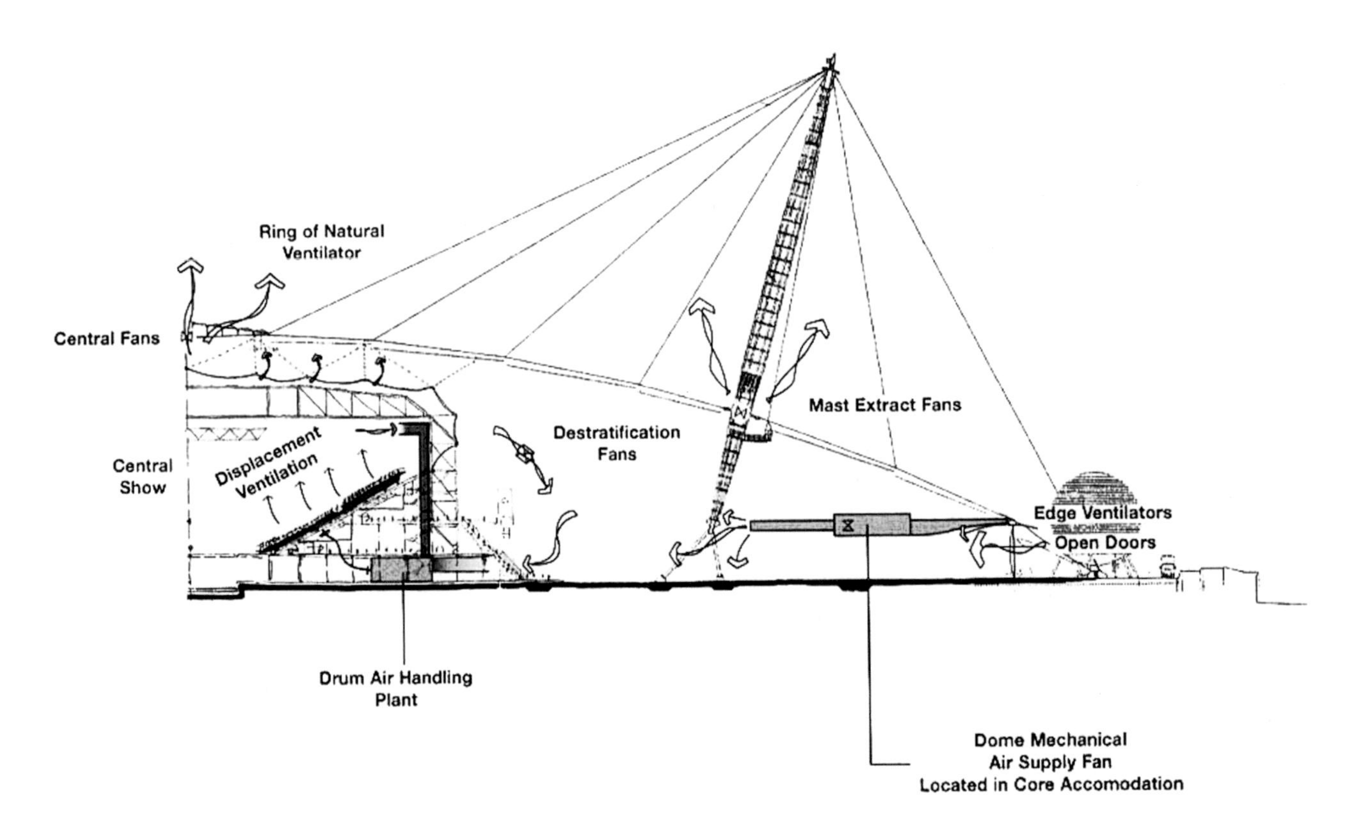

Diagram showing the Dome's ventilation strategy

Image on previous page: Central tension ring to the Millennium Dome

To give as much daylight as possible to offices that look out onto the new atrium, plainly the roof had to be translucent, if not transparent. PVC-covered polyester fabric was chosen because it transmitted light yet was not heavy – a glass roof would have been six times heavier and have imposed unacceptable loads on the refurbished Edwardian buildings. Steel lattice struts span across the void to form the basis for the roof's structure. They support flying aluminium masts that terminate in specially made mushroom nodes which support the fabric. Careful tailoring ensured that fabric, struts and cables form a stable structure which floats over the atrium and over the rear block, allowing a gallery to be formed at original roof level. At the end of the atrium, the fabric is attached to a triangular lattice truss designed to take bending stresses, and the whole structure is designed to wind and snow loads of 600 kilograms per square metre, which has clearly proved adequate in the 17 years since the work was finished. The strategy of using a translucent roof has been a great success, showing that relatively simple move can transform a dreary narrow alley into a luminous semi-public space.

The Dome's 12 masts under construction supported by temporary guys

MILLENNIUM DOME

Location: London, UK
Date: 1999
Architect: Richard Rogers Partnership
Client: The New Millennium Experience Company Ltd
Main contractor: McAlpine/Laing joint venture
Buro Happold services: Structural engineering, long-span and lightweight structures, façade engineering, planning supervision, site supervision, quantity surveying, geotechnical engineering, waste water solutions, site infrastructure, building services, fire safety design, risk assessment, project management

The Dome was built to house a series of exhibition spaces and core buildings. The roof of the Dome is a huge cable net, 320 metres in diameter and clad in 80,000 square metres of tensioned PTFE-coated glass fibre fabric. It is an innovative feat of engineering, yet simple in concept. The roof surface is shaped like a spherical cap.

Twelve 90-metre-high steel masts extend from the roof to support a tensioned net of steel cables, arranged radially on the surface of the Dome and held in place by hangar and tie-down cables at 25-metre intervals. An inner layer of covering reduces thermal gain and improves thermal and acoustic performance. The primary electrical and mechanical plant is housed in the 12 steel cylinders that surround the dome.

In contrast to the relatively small Imagination refurbishment, the £384 million Millennium Dome, designed in London's Docklands with the Richard Rogers Partnership, as a celebration of the second millennium of our era, is one of the largest tensile structures in the world. That the Dome has had a sad history since it was inaugurated is a result of political and entrepreneurial incompetence, and certainly does not reflect on the originality and excellence of the engineering.

The primary supports of the cable net structure are 12 masts, each 90 metres tall. Steel support pyramids anchor the base of each mast and convey loads by piles down to stable strata (the whole area is deeply covered in river silt). Hanging galvanised steel cables attached to the tops of the masts support the net structure, which is formed of 72 pairs of radial stringer cables. Circumferential cables forming a 320 metre diameter ring keep the radial cables in place. To resist uplift forces, screw-in and grouted anchors tie down cables that connect to the hanger cables at 25-metre-intervals round the perimeter.

The patterned membrane segments are tensioned between radial cables, and the cables are separated from the stabilising hoop cables above the membrane surface by steel wishbone fittings in order to avoid snowponding.

For all its apparently elemental shape, the dome structure is complex and radically innovative. The surface of the PTFE (Teflon)-coated woven glass fibre fabric of the tent covering is conceived as a spherical cap. Between the cables, the fabric forms flat panels, and cables were prestressed to stiffen them against imposed loading deflections. Careful detailing was necessary to avoid rain- and snowponding on the fabric behind dams created by the cables, so wishbone nodes that connect cables and fabric were devised to keep cables off the fabric and to act as rotational joints in the radial cables. Such joints (incorporated in the wishbone nodes) were needed because the radial cables would be very long (150 metres) if made in one piece, and flexing at the nodal points would tend to cause the cables to fail by fatigue. So cables had to be made very accurately in discrete pieces to span between nodes. A criss-cross pattern of secondary cables joining nodes serves to stiffen the circumferential structure and prevent progressive collapse if one of the fabric panels should fail or a major asymmetric load were to be imposed. An inner skin of permeable glass fibre fabric (with threads coated with Teflon) was required to minimise the risk of condensation; it is held apart from the outer layer by battens.

View of the finished Dome from across the Thames

The masts are formed from eight 323 millimetre diameter steel tubes braced with rings 2.5 metres apart. Cables are connected to rings at the top and bottom of the masts, and each mast bears on its supporting pyramid through a rubber pot to allow for a degree of rotation. Erection was complicated, for the masts had to have temporary stays, with stresses carefully analysed at all stages to prevent distortion of both masts and cable ring. The fabric had to be prefabricated to very precise tolerances. The considerable ingenuity and innovation achieved in the construction of the Dome was recognised when Ian Liddell, Paul Westbury and their team received the MacRobert Award 2000 from the Royal Academy of Engineering. This was the first civil engineering project to receive this award since the First Severn Bridge some 30 years earlier.

In part, the development of this lightweight structure's vocabulary owes much to the practice's work in the early years with Nick Goldsmith and Todd Dalland of Future Tents Ltd (FTL) on such projects as the Pier Six Pavilion in Baltimore, Boston Harbour Lights and the Bronx Zoo Aviary.

On quite the other end of the scale of magnitude, the pPod, a puppet theatre for the Horse and Bamboo Company, is intended to accommodate audiences of up to 35 people. It travels from town to town on the back of a lorry and can be erected by members of the company in 90 minutes. Designed by Florian Förster and Ian Leaper of Buro Happold with Martin Ostermann of the Berlin firm, magma architecture, the little tent relies on six rectangular frames of aluminium tubes and a skin of PVC-coated polyester sheeting. Attached to the aluminium floor perimeter, the frames are twisted along an imaginary axis. As a result, the tight outer skin is distorted into doubly-curved planes without using any curved elements in the primary skeleton. The structure gains longitudinal stability, and wind forces are resisted by the double curvature of the wall.

Once this outer structure is completed, ply panels can be assembled within the 5 × 7 metre floor perimeter frame. Curved aluminium tubes are erected within the rectangular frames to carry an inner opaque layer of fabric (which is needed, as at the Millennium Dome, to obviate condensation and provide a layer of insulating air). The curved inner aluminium members provide lateral stability to the whole structure. The red outer skin is translucent – transparent even – because it is perforated with microscopic holes that are too small in relation to the surface tension of raindrops to allow them to penetrate. Yet the holes allow glimpses of the inner tubular structure – another instance of slightly magical behaviour by this strange little stage that, in its extremely modest way, shows the kind of mastery of geometry and materials exhibited on a vastly larger scale in the great Dome.

HORSE AND BAMBOO THEATRE

Location: Mobile
Date: 2006
Architect: magma architecture
Client: Horse and Bamboo Theatre
Buro Happold services: Co-designers

The pPod is a unique portable theatre, an innovative structure created by magma architecture in Berlin. It offers short pieces of original, family-friendly, miniature visual theatre and is a dazzling experience that functions as an architectural object and as a performance/exhibition space. Invoked by the form of a kaleidoscope, the form of the theatre revolves around an imaginary axis. Two overlapping skins merge to give an elusive presence and an unexpected experience of the event.

The exterior skin forms a twisted magic box that veils the arching interior performance space. The red exterior is a porous fabric that provides limited views of the internal structure. The aluminium laminate of the interior skin reflects light behind the outer skin. It is opaque on the inside and creates a dark interior for the theatre lighting.

The assembled theatre emits an enchanting glow by night

As at the Imagination building, the new roof of the Dresden railway station demonstrates how fabric structures can brilliantly enhance existing buildings. The present plan of Dresden Hauptbahnhof (which has both through-traffic and terminus platforms) was formed by Ernst Giese and Paul Weinder in 1898, but the building was severely damaged in the Second World War. Subsequent repairs were botched for lack of money, yet most of the fine 19th century metal structure remained and, in 1997, Foster + Partners were appointed to restore the building and replace the old dark roofing with a translucent fabric cover. Buro Happold was made consultants for the fabric skin, with Schmitt Stumpf Frühauf as general structural engineers. A careful study of Teflon-coated glass fibre fabric showed that it was resistant to sparks and diesel fumes and performed well under changing temperatures.

Internal view

DRESDEN MAIN RAILWAY STATION REFURBISHMENT

Location: Dresden, Germany
Date: 2006
Architect: Foster + Partners
Client: Deutsche Bahn Station & Service AG, Regionalbüro für Großprojekte Dresden
Main contractor: ARGE Dywidag und Heitkamp, Dresden
Buro Happold services: Structural engineering, membrane roof (with Happold Ingenieurbüro GmbH, Berlin) and German engineers Schmitt Stumpf Frühauf

The refurbishment of the station hall of Dresden railway station is unique. Sir Norman Foster's design introduces a sculptured fabric roof to cover the existing steel arches of the 100-year-old station. The size of the task, the use of a modern material and the interface between new and old was a challenge for both the client and the design team.

The membrane roof is designed as single panels spanning between the arches. The fabric is fixed to the secondary steelwork, which lifts the line of the fabric above the top chords of the arches and avoids the membrane clashing with the purlins. In the zones between the central and side halls, the membrane roof is pulled down at every second arch to make low points of the fabric structure, so creating the necessary structural form as well as providing rainwater drainage points.

The elegant 19th century arches of the train shed were restored and preserved. They are complemented by new steel transfer structures that allow the fabric to transfer loads elastically to the top chords of the arches. Rooflights supported on this secondary steelwork add to the luminance of the interior and allow smoke extraction and natural ventilation. (The fabric itself transmits 15 per cent of natural light.) Introduction of the new double-curved load-bearing skin makes longitudinal metal purlins largely unnecessary. But the arches remain fragile to horizontal longitudinal thrusts, so such forces are transferred to end bays that are braced to act as 10-metre-wide trusses. So the whole length of the station acts as a complex single structure that can flex slightly with temperature variations. The original volume of Dresden railway station is restored, and the 19th century structure is again revealed in natural light.

Existing roof steel arches

View of sculptured fabric roof

The British Museum offered an even more complex opportunity to relate old and new. Shells share some geometrical properties with domes and have long been another preoccupation of Buro Happold. The Queen Elizabeth II Great Court of the British Museum is the most dramatic Buro Happold shell created so far. Again designed with Foster + Partners, the huge structure seems at first to be remarkably simple, but it is far from so. The site is very curious. The original British Museum, a rigorous and elegant neo-Greek composition by Robert Smirke, was begun in 1824 as four wings round a central courtyard. In 1854, Sydney Smirke was asked to complete his elder brother's work and to fill Robert's court with the British Museum library. He did this by building his famous circular reading room (later used by such disparate authors as Marx and Kipling) in the middle of the court and surrounding it with stores and bookstacks.

Opening the new British Library at St Pancras allowed Robert's original court surrounding the library to be emptied of stacks and his elegant courtyard elevations rediscovered. Foster proposed glazing the whole court, making it the largest enclosed top-lit space in Europe. The stratagem changed the circulation of the museum by making the court the central focus and obviating the often tedious and crowded peripheral routes necessitated by the rectangular plan.

Internal view showing glass and steel roof

QUEEN ELIZABETH II GREAT COURT AT THE BRITISH MUSEUM

Location: London, UK
Date: 2000
Architect: Foster + Partners
Client: Trustees of the British Museum
Buro Happold services: Structural engineering, building services, fire engineering, geotechnical engineering, planning supervision

Designed in 1823 by Robert Smirke, the British Museum's Georgian buildings originally consisted of four wings containing galleries set around a large rectangular courtyard, which was later filled with extensions, including the circular reading room by Smirke's brother Sydney. This is now the centrepiece of the largest covered courtyard in Europe, enclosed by a spectacular glass and steel roof.

The groundbreaking lattice roof which covers the central courtyard of the museum is made up of intricate steel and glass latticework which creates a delicate and unobtrusive canopy with no supporting columns. It is a highly practical solution, allowing maximum natural daylight without solar glare, through the use of neutral-tint fritted glass.

The move is bold, and its engineering implications manifold. For a start, the reading room is not actually in the middle of the courtyard but placed eccentrically 5 metres towards its north side. Though the heights of the buildings appear to be the same, they are not. The reading room is a very fragile shell on weak foundations, a brick drum stiffened with iron framing. To strengthen the cylinder, 20 columns (457-millimetre steel tubes filled with concrete) were arranged round the reading room so that they can carry roof loads down to foundations. The columns are hidden behind the new stone cladding of the drum (which was never intended to be seen from outside in the first place), and they are crowned with a snow gallery that collects the vertical loads and acts as a stiff diaphragm to prevent the roof spreading laterally over the drum.

Creating the 6-metre-deep basement was a complicated problem in itself. Laboratory testing of the soils showed that jet grouting could be used to underpin Sydney's structure and to form lateral restraint. During construction, movements were continuously monitored, and procedures were varied when deviations from calculated positions threatened. In the end, the vertical movement of the reading room agreed with its predicted amount, and lateral displacement differed by 3 millimetres.

Every piece of steel and glass is a different size.

Nodal connection of rectangular tube structural members

The steel and glass shell roof is formed as a toroid to facilitate easy transition between inner circle and outer rectangle – and to ensure that the new roof is not visible behind Robert Smirke's magnificent south front (a planning requirement). Round the outer perimeter, sliding bearings take roof weight down to the masonry walls that are strengthened with steel to take vertical loads but could not take horizontal thrusts. The roof lattice consists of 6,000 individual members that had to fit together exactly, so each was manufactured with great precision from grade D steel (rarely used in buildings and more often to be found in marine or petro-chemical work). Radial elements span out from the reading room, and are connected by two opposing spirals to form a lattice, hence every piece of steel and glass is differently sized.

Controlling the temperature of such a large glass-roofed area is clearly difficult. Many different climate control techniques are needed to cope with the large range of internal conditions, from the restaurant on top of the drum directly below the glass roof, to the basement lecture theatres. To control solar gain, all glazing is double and fritted. Summertime heat build-up is countered by integrated design of perimeter displacement ventilation, natural ventilation and some mechanical extract. Daytime air conditioning chillers are used at night to reduce the temperature of the slab of the Great Court, adding a flywheel cooling effect during the day. Four separate plant rooms are required to cope with the different conditions.

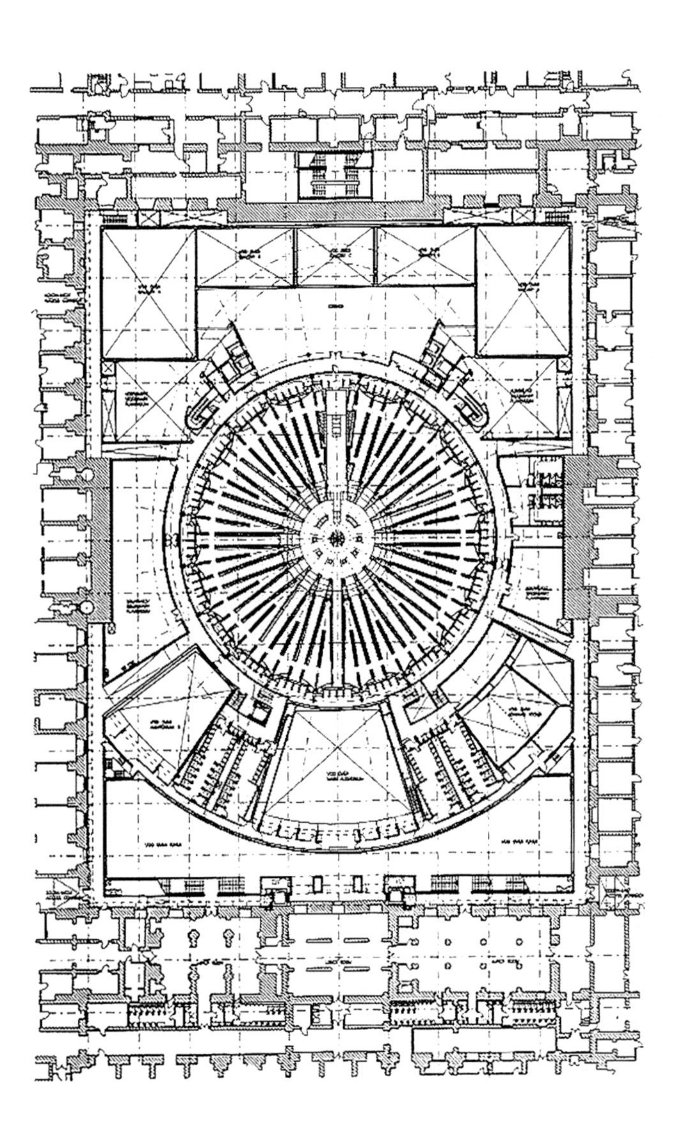

Plan showing foundations to reading room and original courtyard structures

If the British Museum shell was required by the planners to disappear when seen from ground level, the shell of the Sage Music Centre was necessarily very visible. As part of the regeneration programme for Gateshead, previously a dilapidated industrial city on the river Tyne, the centre provides a major public performance space and the home of the North of England Symphony Orchestra. As well as the main auditorium, the complex has a smaller performance hall and a rehearsal space, each acoustically separate, but all linked by terraces and esplanades; the sub- and superstructures of these elements were designed by Foster + Partners with engineers Mott MacDonald. All had to be enclosed in an outer envelope structurally separate from that of the halls to ensure acoustic isolation. Foster + Partners held an ideas competition among engineers for a solution.

Buro Happold won. The proposal was based on a tubular triple wave form that reflects the volumes inside. Structurally, primary trusses are built in the valleys. Spanning between these arches are secondary arched members so that the whole becomes a complex interlocking toroidal shape. Tertiary strut and tie members lock the secondary elements into place to make the whole roof act as a shell and to limit deflections.

SAGE MUSIC CENTRE

Location: Gateshead, UK
Date: 2004
Architect: Foster + Partners
Client: Gateshead Council
Buro Happold services: Structural engineering (roof only)
Principal engineer: Mott MacDonald

This project consists of three separate buildings: two main acoustically excellent performance halls and a third smaller rehearsal space. These buildings are interlinked with terraces, esplanades and basements, and are covered by a single roof. Buro Happold based their idea for the roof on a structural triple wave form in the long direction and a single closing arc form in the short direction; this related to the architecture and helped define the form of the three music halls beneath. As the form was potentially so complex, ways had to be found to construct the roof both easily and simply to achieve the tight budget constraints. As pure shells could not be built easily on the site to budget, a more hierarchical form of structure was adopted.

Internal view of offset connection from horizontal curved roof beams to vertical glazing rails

External view showing entrance and glazed roof

Fully-glazed end walls are offset by the undulating shell. This is covered in a rainscreen of interlocking stainless-steel panels above a weather-tight membrane and insulation carried by profiled metal sheeting supported by the secondary arch members. Internal gutters in the void between membrane and rainscreen remove interstitial water and allow the outer surface to be smooth, so creating one of the most dramatic forms on the lower river, architecturally now one of the country's most lively areas.

View of hospital's atrium with external shading and springing point of V steel tubes connecting to diagonal grid roof

Internal view of steel diagonal grid roof and glazing above looking towards the new hospital wing

A major improvement to the dreary cityscape of London's Lambeth Palace Road is the great glass barrel vault of the Evelina Children's Hospital, designed with Hopkins Architects. The hospital is intended to be the main paediatric centre for the capital and so needs the large protected recreation area enclosed by the roof where children can play all year round. The building's basic structure is a seven-storey flat slab concrete frame with the glazed steel diagrid vault made of circular hollow sections of varying wall thickness. The roof is four storeys high, springs from three floors of more conventional accommodation and curves up and over level seven. Diagrid nodes have to be capable of responding to thrusts in vertical and two horizontal directions caused by wind, snow and thermal movement. As a result, structural deflections are limited to what can be accepted by the glazing system. The lower part of the big space has vertical glazing that meets the roof structure with a hinge joint that allows both vertical and horizontal movement. Inside, the big, sunlit well-planted space is enlivened by scarlet lift towers with constantly moving cars. It has a play area, a school for long-term patients, and a restaurant. Its floor is pierced by voids to allow natural light down to the out-patients and specialist treatment areas below. Designed as a response to children's dislike of 'long scary corridors', the conservatory has proved popular, and acts as the social centre of the whole building.

EVELINA CHILDREN'S HOSPITAL

Location: London, UK
Date: 2003
Architect: Hopkins Architects
Client: Guy's & St Thomas' Hospital Trust
Buro Happold services: Structural engineering, fire engineering, planning supervision

The brief was for a landmark building, exciting, fun and friendly, yet with a design necessarily innovative to deliver an efficient, cost-effective and functional hospital. Evelina Children's Hospital is a seven-storey concrete frame building with a 120-bed integrated children's hospital facility. Its purpose is to centralise under one roof all the paediatric functions that were once spread throughout London. The design philosophy was to create an exciting environment for children with the large open space formed by the conservatory roof serving as a focal communal area to the hospital. There are many features such as a large transfer truss, scenic lifts and a steel diagrid roof as well as a link bridge connecting Evelina to the adjoining Guy's and St Thomas' hospital.

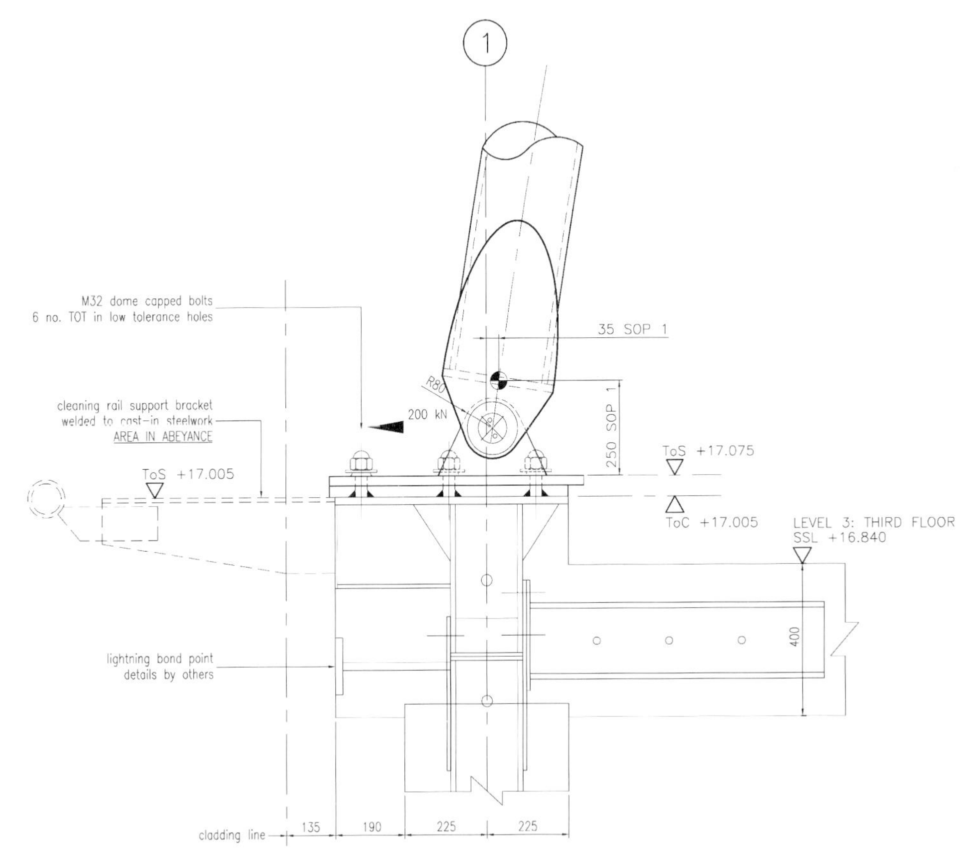

Detail of V tube springing point connecting to diagonal grid roof

Interior of workshop showing natural lighting

WEALD & DOWNLAND OPEN AIR MUSEUM

Location: Chichester, Sussex, UK
Date: 2002
Architect: Edward Cullinan Architects
Client: Weald & Downland Open Air Museum
Buro Happold services: Structural engineering, building services

The Weald & Downland Open Air Museum is the leading museum of historic buildings in England, designated by the Government for the pre-eminence of its collections and its building restoration work, especially timber framing. The new two-storey Downland gridshell provides open access to the public to view the museum's work. The lower level contains an archive store and conservation workshops. The upper one houses the timber framing workshop beneath the innovative new timber gridshell roof. This greenwood structure uses new techniques researched especially for this project. It was funded by the National Heritage Lottery.

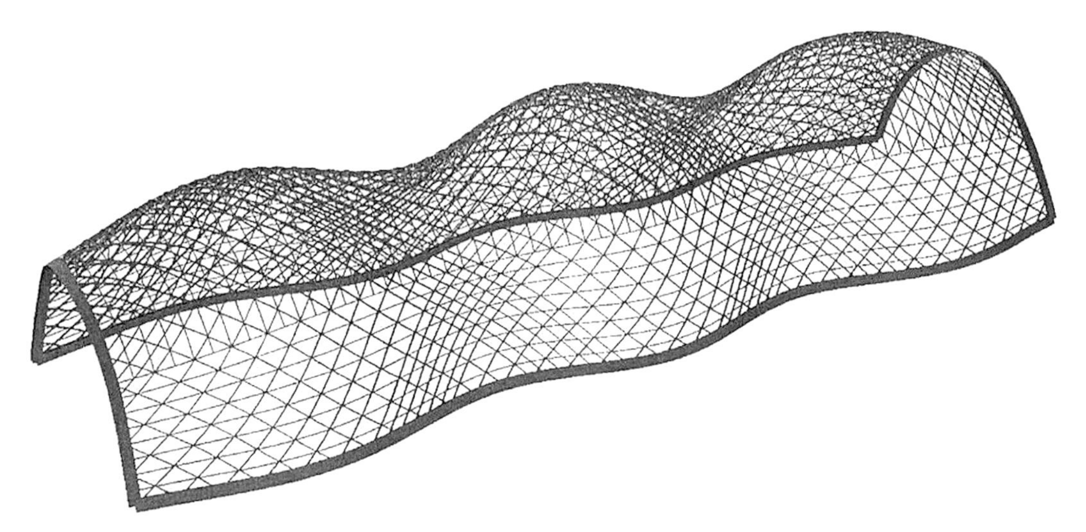

Model of final form of double-layer gridshell on its foundations and end arches

Image on previous page: External view of the Weald & Downland Open Air Museum gridshell under construction

Particularly in the last few years, gridshells have been one of Buro Happold's major research interests, though Ted Happold, Ian Liddell, Terry Ealey and Michael Dickson worked with Frei Otto on some of the earliest ones in the 1970s, such as the multi-purpose pavilion for the German Federal Garden Exhibition at Mannheim (1975), with architects Carlfried Mutschler and the brothers Joachim and Winfried Langner. Like egg or mollusc shells, gridshells have very high strength-to-weight ratios but, unlike natural structures, their continuously curving forms are composed from a grid of slender lightweight elements that are assembled into curves during construction. Capable of spanning surprising distances and rising to considerable heights, gridshells were until recently difficult to analyse mathematically and hard to construct.

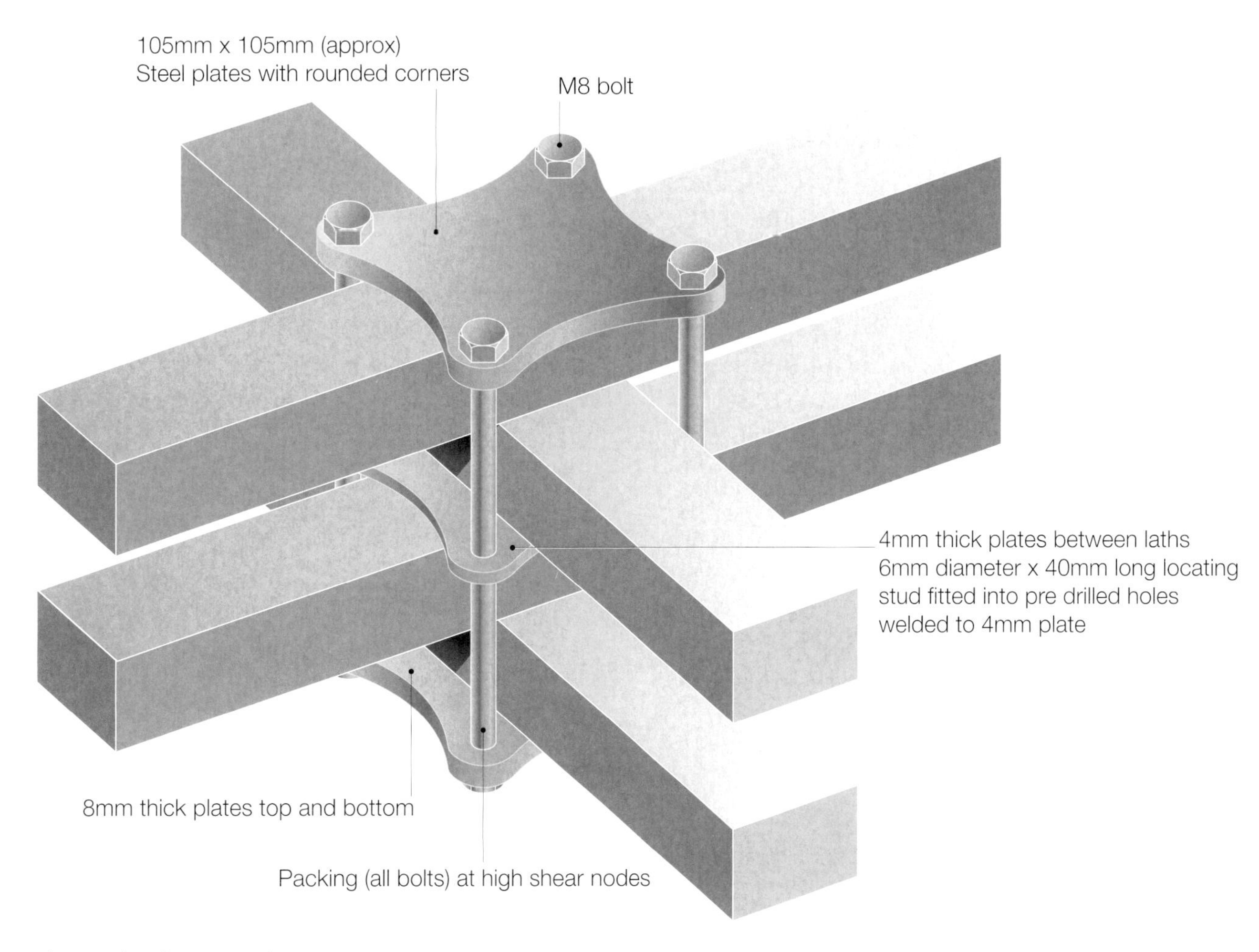

Diagram showing patented node enabling connection of triangulating fifth layer of laths

In many ways, the workshop for the Weald & Downland Open Air Museum at Chichester in Sussex, England, is a precursor of recent work. A leading international centre for conservation and study of historic buildings, the museum is in a 19th century landscaped park on the flanks of the South Downs. Its collection includes some 50 rescued local buildings that date from the 15th to the 19th centuries. Most of them are timber framed (half timbered) and the museum was concerned to complement the work of previous generations by making a modern design as ingenious and reliant on local materials as those of Downland builders centuries ago. Edward Cullinan Architects were chosen as they had a fine track record of innovatory work with old buildings that respected the past by sympathetically adding to it, rather than copying. Cullinan asked Buro Happold to be structural engineers, for the two firms had worked together on experimental buildings using forest thinnings of greenwood at Hooke Park College in Dorset.

Western red cedar
cladding of gridshell

Downland Museum's brief was to make a large open workshop for treating and repairing elements of traditional timber frames and to provide climate-controlled archival space. It is a national facility for the study and practice of building conservation, so large sections of traditional buildings must be treated under cover. In essence, the design was extremely simple, with the archival areas dug into the thermally stable chalk hillside below the clear-span workshop that has a re-used 16th century timber plank floor 80 millimetres thick. The museum has a three-mounded roof, forming what the designers call 'a three-bulb hourglass', 12 to 15 metres wide and 50 metres long. The roof workshop is formed of 50 × 25 millimetre green oak laths made into a double-layered gridshell. The double shell was required because a single one would have required laths with such sturdy sections that they would not flex easily – flexible components were needed to achieve the final form. Oak was chosen not only because it is readily available locally but because it is durable, very strong yet plastic when green.

Patented node as support for cladding and triangulating laths

The three-domed, double-curved gridshell roof viewed from the gardens

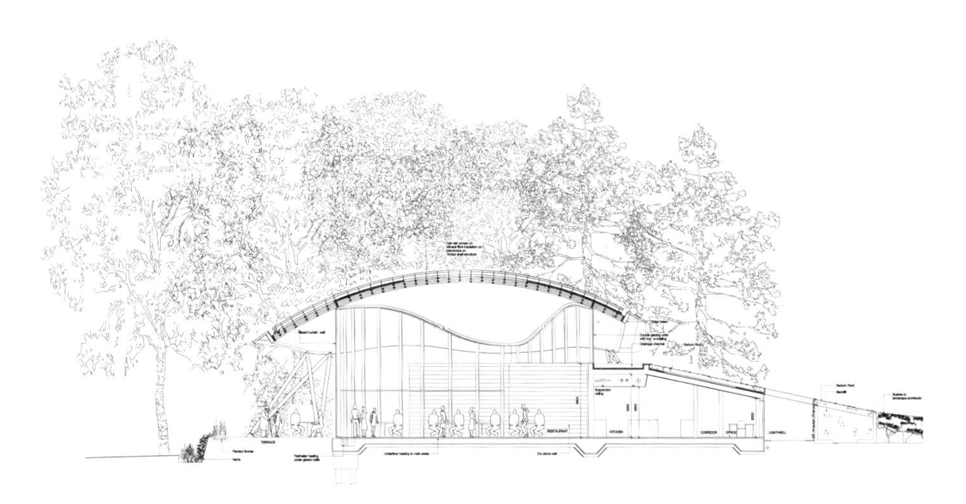

Garden elevation to the visitor centre

Internal view showing the curved timber roof

Erection was innovatory. Gridshells are normally formed into their finished shape from the flat gridmats they start as by pushing them upwards from below or by pulling them upwards with cranes. Such techniques work against gravity, which can cause unpredictable problems in the slender components as they move into their final form and position. At the Downland workshop, the gridmat (made flexible with special patented sliding and rotating joints) was formed on a scaffolding-supported platform 7.5 metres above the workshop floor. Scaffolding was then carefully dismantled in calculated stages, allowing gravity to drape the building into its final curved form, assisted by carpenters using ropes and clamps. The structure was progressively fixed to the wood floor of the workshop and a fifth layer of oak rib laths were applied to the form, longitudinally over the lower two thirds of the building and transversely in the upper third. These elements ensure that the whole structure is triangulated to finally act as a shell. They also serve as cladding mounts for the local western red cedar boarding of the lower part and the polycarbonate and timber clerestory.

With its shaggy cedar exterior gradually changing in exposed areas from brown to grey, topped by shining bands of clerestories, the new workshop happily takes its place amid old buildings so carefully collected on the slopes of the Downs. The Sussex shell demonstrates how to economise in resources, how to do more with less and how to use elements (thin oak laths) that were previously unconsidered as structural elements.

SAVILL BUILDING

Location: Windsor, UK
Date: 2006
Architect: Glenn Howells Architects
Client: Crown Estate
Design, construction, contractor for gridshell: Green Oak Carpentry
Consulting engineers: HRW (Haskins Robins Waters)
Buro Happold services: Structural engineering of roof

Situated on the edge of Windsor Great Park, the Savill Building highlights the potential of lightweight timber frames, and the dramatic curved timber roof provides a major new landmark attraction. This elegantly engineered, environmentally friendly building is made entirely from timber harvested from the adjacent Crown Estate, and is the largest timber gridshell structure in the UK. The three-domed, double-curved sinusoidal shape is clad in oak, with the shell perimeter comprising a tubular steel beam supported on steel quadripods.

The roof is 90 metres long and is formed using a delicate interlocking timber lath construction of larch, shaped and jointed by specialist carpenters working to a unique computer-generated design.

Edge detail of timber roof

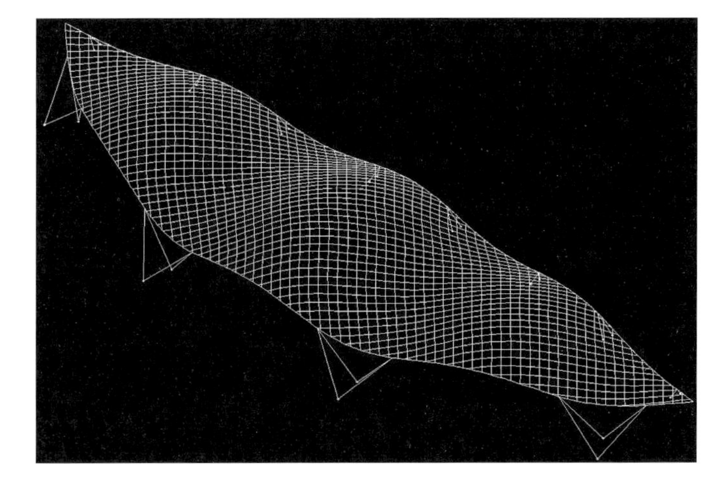

Computer model of timber roof structure and steel tubes supporting ring and cables

Following the Downland design, the Japan Pavilion for the Hanover 2000 Expo was one of Buro Happold's most radical gridshell experiments. Shigeru Ban, the architect, had already made his name as an ecologically aware designer and had demonstrated the potential of paper and cardboard in structures like his church in Kobe, quickly built by its own congregation after the city's 1995 earthquake. He was naturally keen to explore further the potential of cardboard tubes in building construction, and was an ideal choice for designer of a temporary pavilion for an ecologically orientated expo. Frei Otto was made consultant to the project and introduced Buro Happold as structural engineers.

Initially, the design was for an amphora-shaped exhibition hall 72 metres long with a maximum span of 35 metres: it was to be a doubly-curved shell made of a grid of paper tubes, but tests and calculations showed that solution to be infeasible. To reduce the diameter of the cardboard tubes, the overall shape was modified into a three-mounded form as at Downland, and timber ladder elements were added. So the basic metre square diagonal grid of 120 millimetre diameter cardboard tubes was supplemented by the laminated timber ladder-like arches spanning the width of the building at 3 metre centres. Stiffness was added to the ladder arches by chorded thin stainless-steel cables. At right angles to the ladder arches were 60 × 95 millimetre laminated timber purlins. Cladding was fixed to the timber elements of structure and was of two kinds: between the ladder arches, there was a double membrane of paper with a protective cover of polymer-coated polyester; under the arches were single layers of clear polyester membrane that acted as gutters. Clear and translucent claddings acted together without glare to create an interior filled with gentle amber light.

Interior of the finished pavilion showing the paper tube gridshell overlain by strengthening timber-tied arches

JAPAN PAVILION

Location: Hanover, Germany
Date: 2000
Architect: Shigeru Ban
Client: Japan External Trade Organisation
Buro Happold services: Structural engineering, sustainability design, materials

The award-winning Japan Pavilion was commissioned and built for the Expo 2000 trade fair in Hanover, an international showcase for the latest in technological and environmental innovation from around the world. Commissioned by the Japan External Trade Organisation, the pavilion was developed as a double-curved gridshell to span the exhibition space column-free. The structure of the 35-metre clear-span enclosure consisted of a core of paper-covered cardboard tubes connected to a series of timber arches stiffened by cables. Over 80 per cent of these construction materials were recycled after the exhibition closed.

One of the emphases of the brief was to make as little permanent impact on the site as possible, so concrete foundations had to be avoided. Both timber ladders and paper tubes were connected to timberboard-clad steel A-frames along their perimeter. These A-frames, laid directly on the ground, were weighed down by being filled with sand to obviate for foundations in the ground. Construction began by laying out the cardboard tubes in their grid on a temporary scaffold, which was gradually pushed up by jacks into the final shape of the finished building over a three-week period. Joints between tubes were made with fabric bands that allowed the whole structure to flex as it moved into position. Work began in the middle of the building and then expanded simultaneously towards each end. Once the timber elements of the central section were in place, lacing of the stainless-steel cables could start and the membranes followed immediately afterwards.

In many ways, the Japanese pavilion was the star of the Hanover show. All expo pavilions aim to cut a dash, but the Ban/Buro Happold/Otto contribution was not just an innovatory form and a memorable space, but a notable lesson in recyclability, for almost all its elements, from the offices (which were in commercial containers) to the cardboard and paper structure and fabric, could be re-used – and were.

In some ways this theme of recyclability was exploited to its ultimate by the Nomadic Museum for Pier 10 in the Hudson River in New York by Shigeru Ban and Buro Happold's New York office. The exhibition theme, based on the form of the Arsenale Venice, hails in the painted ro-ro boxes that form the side walls to a translucent roof that is itself supported by large-diameter paper tube columns.

Paper tube gridshell
during construction

Visualisation of the canopy

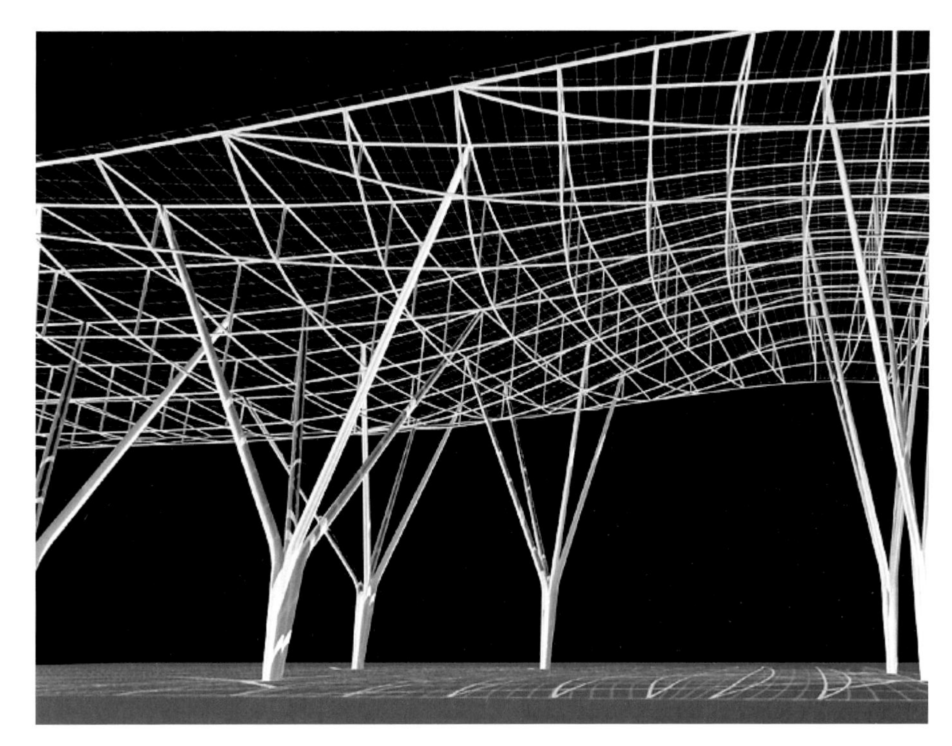

3-D model of structure

ROPPONGI CANOPY

Location: Tokyo, Japan
Date: 2000
Architect: Skidmore Owings & Merrill
Client: Roppongi Midtown Project
Buro Happold services: Lightweight and long-span structures, materials

This stunning canopy structure, an architectural achievement in its own right, provides a memorable entrance to a large development – a plaza area with three buildings – in the Roppongi area of Tokyo. The canopy, which spans escalators linking Gaien Higashi Street with an underground mall, provides protection from wind and rain and creates a unifying link. The canopy covers an area of about 30 × 80 metres.

The challenging layout of the glazing grid was generated using complex geometry, with the condition that it should be constructed with flat, reasonably sized panels of glass so that the structural grid would allow for maximum transparency and lightness. Buro Happold used SmartForm to analyse these complex surfaces and suggest principal directions for arches or structural reinforcements in order to fit a set of four-sided flat panels to a curved surface.

The Green Oak Carpentry Company, which built the Downland shell, asked for Buro Happold as consultants when they were appointed to be sub-contractors for the roof of Glenn Howells Architects' visitor centre at the Savill Gardens in Windsor Great Park (Haskins Robinson Waters were the structural engineers for the rest of the building). The Windsor building has similarities with the Sussex and Hanover ones: for instance, it has three mounds in the roof and a lath structure. But there are some key differences: Howells wanted to restrict the height of the building to control its impact on the gentle landscape; the four-layer shell structure does not form both roof and walls, but is supported at eaves height by a tubular steel ring beam; and the green laths are of larch not oak. All the wood used in the roof, including its oak rainscreen, comes from selective felling in the commercially managed (but very ecologically conscious) Windsor Forest which is run by the Crown Estate.

The building is approached through a grassy bank that conceals plant rooms, offices, a lecture theatre and even a top-lit greenhouse. So when you get inside, the grand undulating sweep of the highly stressed relatively flat roof and its delicate grid is unexpected, and so is the fine view from the glass wall that overlooks the gardens. On that side, the roof projects over the glass to form a sheltered arcade articulated by the tubular steel props of the ring beam. At Windsor, the grid supports a thin layer of birch ply that both acts structurally and forms a light-reflecting ceiling. The building is important because it shows that gridshells are not just to be used to make separate one-off buildings, but their virtues can be incorporated in more conventional structures that can become parts of urban and landscape form.

External view of the structure

View showing individual cardboard roof panels

WESTBOROUGH PRIMARY SCHOOL

Location: Westcliff-on-Sea, Essex, UK
Date: 2001
Architect: Cottrell and Vermeulen Architecture
Client: DETR
Buro Happold services: Structural engineering, building services, sustainability consulting

An intensive research study into the viability of cardboard as a construction material by the Buro Happold RDI (Research, Development and Innovation) team led to a project to design a building from the material for Westborough Primary School, as an after-school club consisting of a changing area, kitchenette and toilet block. Being made from waste paper, cardboard has the potential to be a very 'green' building material. Cardboard was used in structural tubes to support the roof, structural panels principally to stiffen the timber frame, as insulation for the walls and roof and also as surface layers.

Water poses the greatest threat to cardboard, with fire close behind. In designing the building, Buro Happold found successful solutions to these technical challenges: the building still stands and is repeatedly the subject of scientific and educational TV programmes.

Interest in the potential of cardboard in buildings was re-awakened in a small after-school club building for Westborough Primary School in Westcliff-on-Sea near Southend, Essex. Here, the architects Cottrell and Vermeulen, were concerned to reduce the environmental impact of their building and to use recycled and recyclable materials as much as possible. Working with manufacturers, the design team evolved a new form of stressed-skin cardboard panel that has robust structural properties as well as acoustic and thermally insulative ones. It is a sandwich in which three 50-millimetre layers of honeycomb cardboard separated by 2 millimetres of solid card are contained between two 6-millimetre layers of solid cardboard and edged with wood (to facilitate jointing).

Cardboard quickly goes soggy when wet, so the outer surfaces of the panels are covered in a waterproof breathing membrane that can be removed when the components are recycled. The other commonly raised objection to using cardboard in buildings is that it burns. Studies showed that the solid card used in the school burns no more dangerously than timber – slowly and by charring. Panels were treated with a flame retardant to further inhibit flame-spread, so in a single-storey building like the school club, occupants would have ample time to escape. Roof loads of the little building are carried on standard cardboard tubes like the ones used by Ban at Kobe. Manufacturing limitations of the cardboard panels determined that the middle of the roof has to be supported by a timber truss – a compromise reluctantly accepted by the design team. But there is hope that the system can be adapted for multi-storey work, as well as much larger buildings. If so, cardboard manufacturing problems will surely be overcome, making the material a major component in ecologically conscious construction.

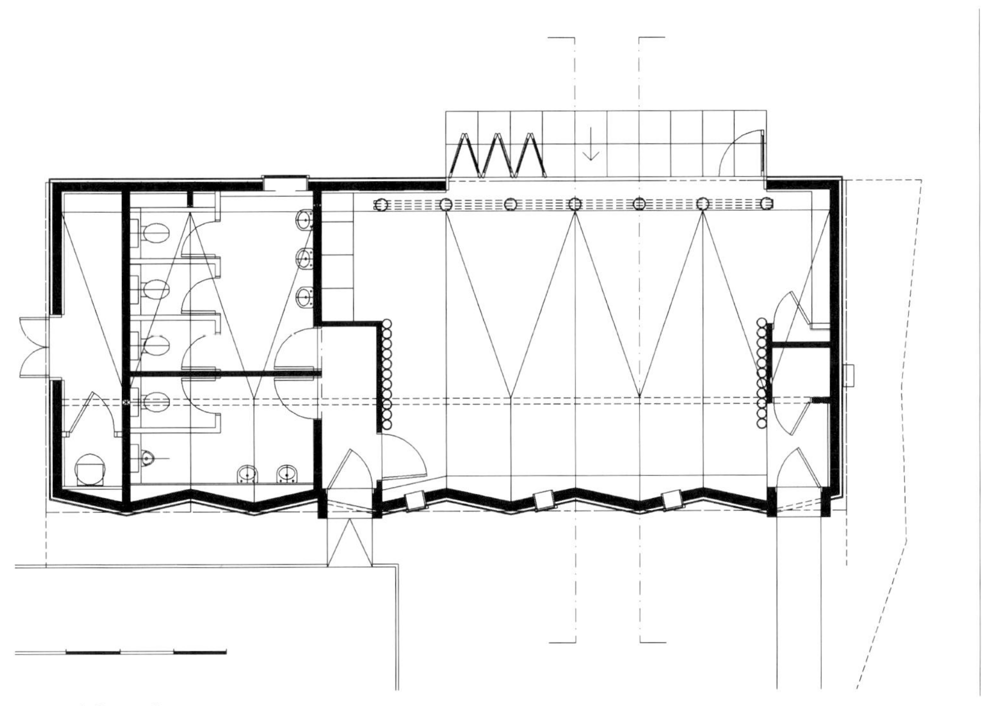

Ground floor plan

External view

Internal view of the auditorium

Image on previous page: External view of Winter Garden, part of Sheffield's Heart of the City Project

PERTH CONCERT HALL

Location: Perth, Scotland, UK
Date: 2005
Architect: Building Design Partnership
Client: Perth and Kinross Council
Buro Happold services: Building services, structural engineering, fire engineering, environmental computational fluid dynamics, strategic IT

This important public sector project gives the city of Perth a fine new concert hall that doubles as a conference venue. Buro Happold was asked to provide a range of multi-disciplinary engineering services for the building of the hall. Its main auditorium is able to seat an audience of 1,400. For maximum comfort it was important to ensure that ventilation would work efficiently while at the same time ensuring a low-energy design solution for the concert hall. For this reason Computational Fluid Dynamics (CFD) modelling was used. This helped simulate natural ventilation concepts within the venue's foyer. The displacement ventilation system within the auditorium itself was modelled.

Over half the human population of the planet now lives in cities, and the proportion is likely to continue to increase in the foreseeable future. Building in cities is becoming more and more important, and urban form and urban architecture are the focus of much debate. Some architects have decided to respond to what they perceive to be the potentially amorphous and anonymous nature of the city by creating buildings that are obviously strange and are intended to become urban landmarks. One of the simplest ways of making a building look unusual is to subvert the normal rules of structure – or at least make it seem as if gravity is of no consequence. Such buildings make difficult demands on structural engineers, who have to make paradoxical forms work while concealing how the trick is pulled off.
A striking example of this approach is Palestra, a speculative office block designed by Will Alsop to accommodate 3,000 people in Southwark, in the South London's Bankside.

PALESTRA

Location: Southwark, London, UK
Date: 2006
Architect: Alsop Architects
Client: Southpoint General Partner Ltd
Buro Happold services: Structural engineering, building services, ground engineering, fire engineering, façade engineering

The building features a series of quirkily angled perimeter columns that support the structure and a top section over eight storeys which appears to be disjointed and separate from the rest of the building. These design elements pose significant challenges. The brief demanded innovative solutions to ensure that the initial concept of two separate floating boxes could be realised, a further challenge was to provide the number of required floors within the planning envelope to make the project commercially viable. The building's complex geometry sits above part of the Jubilee Line Extension tunnels.

View of glazed cantilevered top section

Structural integration of twin primary cellular beams with building services

Palestra is two differently glazed boxes, one apparently precariously and almost randomly perched on top of the other, which itself is made to seem as if it might fall over. The lower box is nine storeys high and its external walls (though not its interior ones) are set at 11 degrees to the horizontal. Hence, the glass curtain wall of the box touches the ground in the east of the site and slants upwards to form a double-height public entrance space at the west end. Alsop has a fondness for columns set at arbitrary angles to the vertical (what he calls 'dancing columns') and 12 of these prop the perimeter of the entrance space, making it look as though they have been knocked sideways during construction and might collapse at any moment. The top three storeys form the upper box, which overhangs the lower storeys by 1.5 metres on north and south sides. At the west end of the building, the box cantilevers dramatically 7.5 metres to the west over Blackfriars Road, a move necessitated by the need not to overshadow neighbouring residential properties.

External view showing the 'floating' top section

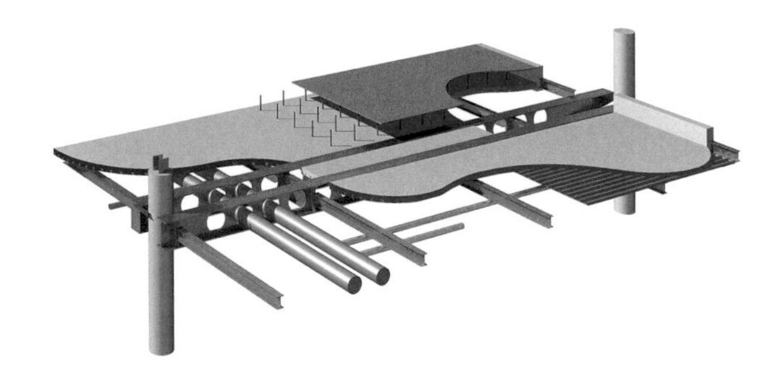

Computer rendering showing integration of twin primary cellular beams (to save depth) with services

The main structure is a steel frame with 508-millimetre circular hollow-section columns supporting twin primary cellular beams that themselves carry secondary beams bearing metal decking with a 140-millimetre thick in-situ concrete slab. The holes through the primary cellular beams allow for passage of longitudinal building services elements. Columns are filled with reinforced concrete, so increasing speed of erection and structural properties; the need for external fire protection is obviated. In the end, this quite orthodox, very stiff K-braced frame comprising stair and service cores absorbs and transfers the eccentric loads imposed by the architecture.

Because of their angles, the dancing columns generate horizontal forces at top and bottom; these vary because column inclinations differ, and the upper forces are taken up and resolved by a fabricated steel transfer structure at second floor level. In this, the columns' unbalanced horizontal forces are balanced against each other. A similar structure has been evolved for the junction of the two boxes between the seventh and ninth floors. Dancing columns are used again to mask the transition between the two forms, and make structural links between the different structural grids (lower box 10 × 7.5 metres, and the upper one 12 × 7.5 metres). Such clashing geometry ensures that, at the perimeter, none of the upper columns can bear directly on a lower one, so the inclined columns are a vital part of the whole structure. The most dramatic move in the whole composition, the great cantilever of the top box, was made possible by plate steel Vierendeel girders that transfer the overturning loads of the cantilever back to the core frame: a heavy solution perhaps but, as diagonal props were not allowed by the architect, it is difficult to see what else could have been done.

View inside atrium

Angled columns that support the structure are organised around the perimeter.

Another architectural approach to urban building is to work with the existing fabric and endeavour to enhance it. Despite the fact that cities are our oldest and most complex artefacts and works of art, many were torn apart in the 20th century by war, primitive planning and transportation measures, changing patterns of production or, in the poorer parts of the world, by the combined effects of industrial and agrarian revolutions that sent millions from the land to traditional cities that are unable to accommodate them in a civilised way.

The Millennium Galleries complex in Sheffield, designed by architects Pringle Richards Sharratt with Buro Happold as environmental consultant, is one of the most successful recent examples of a modern central city building that adds to and interacts with its context. It is part of a plan to rejuvenate the centre of the city, which has radically changed its nature over the last 50 years, as the city's traditional staple industries were transformed. Near the two existing theatres and the 1930s public library, the new complex joins previously disconnected city spaces and streets with an elegant glazed arcade. In so doing, it re-awakens an eminently sensible West Yorkshire tradition. In a comparatively cool and wet climate, it makes sense to shop under cover, and in the centre of nearby Leeds are several fine, well-preserved 19th century arcades.

The Winter Garden structure is made up from laminated timber arches

THE MILLENNIUM GALLERIES AND WINTER GARDEN

Location: Sheffield, UK
Date: 2002
Architect: Pringle Richards Sharratt
Client: Sheffield City Council
Buro Happold services: Structural engineering, building services, ground engineering, fire engineering

With the Millennium Galleries, the Winter Garden forms part of the much bigger redevelopment of Sheffield city centre: the Heart of the City Project. Both are designed to be highly energy-efficient and provide a model for sustainable development in city centre sites. The Winter Garden is a spectacular glazed public space in the heart of the city centre, its structure formed by a series of timber arches of glue-laminated (Glulam) European larch.

With four galleries under one roof, the Millennium Gallery was created and designed to be thoroughly accessible to the disabled and elderly, and incorporates many facilities. The Millennium Gallery is made mainly of glass and white concrete, with marble floors. It is a modern, light and spacious building, and is a much more welcoming space than traditional institutional galleries.

Though the Galleries complex is large in relation to the scale of most of the existing inner city, the architecture is unobtrusive and well-mannered, consisting largely of glass and white precast concrete. All services are kept beneath the marble floors, so both the new covered street (the Avenue) and the vaulted galleries it serves can be simple, clean, airy, almost neo-classical spaces. The Galleries are largely naturally lit, with daylight from clerestories being bounced off reflectors into the vaults from where it softly radiates into the spaces. Artificial luminance is carefully co-ordinated to increase gradually as natural light fades. Exhibitions vary from temporary shows travelling from the major national museums in London to a metalwork gallery demonstrating local skills and the Ruskin Gallery, which contains the delicate, beautiful and idiosyncratic collections given by the great Victorian critic and philosopher to his Guild of St George, founded to educate and illuminate the workingmen of the city. All the different types of exhibit need particular kinds of lighting.

The Galleries are complemented by a café/restaurant, a shop and a landscaped Winter Garden (the latter under a glass and laminated timber arched roof). All these add additional micro-climates to the ones required by the Avenue and the galleries themselves. All are co-ordinated (and their energy needs minimised) by an electronic building management system – essential to control such a varied and multi-orientated urban intervention.

White precast concrete elements of the Millennium Galleries under construction

Internal view

Glulam arches share loads from alternating arches which are carried to the ground by inclined timber struts and stiffened laterally by galvanised steel trays.

The Sheffield Galleries relate old and new parts of the urban fabric in a gently reciprocal way. Sometimes, more radical relationships of ancient and modern are needed. One of the key buildings of central Manchester, another great Victorian industrial city, was the Free Trade Hall, a rather heavy but dignified renaissance palazzo designed by Edward Walters in 1856. Second World War bombing destroyed much of the building, but two façades were saved, and the building was redeveloped in the 1950s as the home of the famous local Hallé Orchestra by Leonard Howett. This incarnation of the hall became redundant when the orchestra moved to a new hall with acoustics appropriate to its international status.

Rear elevation to new hotel room wing

FREE TRADE HALL HOTEL

Location: Manchester, UK
Date: 2004
Architect: Stephenson Bell Architects
Client: Radisson Hotels
Main Contractor: MacAlpine & Laing O'Rourke
Buro Happold services: Structural engineering

For over 150 years, Manchester's well-known Victorian Free Trade Hall was home to the Hallé Orchestra. But the orchestra moved to the new Bridgewater Hall, leaving the famous building empty on the hands of Manchester City Council. Stephenson Bell's winning entry in an open architect/developer competition slots a 263-bed, five-star hotel onto the site, with an additional 1,500 square metres of meeting and function space, two restaurants with bars and a health spa for the entertainment and convention quarter of Manchester.

Planning constraints limited the height of the contemporary block to reduce its impact on the retained Victorian façade, while commercial pressure required increased accommodation. Yet the design constraints have produced a hotel with extremely efficient use of space.

Owned by the city council, the building stood empty in a rather forlorn area of the city while arguments raged about its historic value. It was rescued in the end to become the site of a major hotel, provided that the two façades by Walters were retained. Stephenson Bell won the architect/developer competition for re-using the building and asked Buro Happold to be consulting engineers. A fundamental structural problem of urban hotels as a building type is that the small spans of the bedrooms usually have to be placed over the much wider spans of the public rooms: a problem exacerbated by the need for acoustic insulation – and hence mass – between bedrooms. Architects and engineers resolved the difficulty by putting bedrooms in a 15-storey tower and using the Walters façades to front the public areas and offices, which are composed broadly into a palazzo plan. The two elements are connected by a transparent triangular atrium clad in planar glass. Bedrooms were constructed using in-situ tunnel-form concrete casting techniques that allowed rapid completion of a stiff, acoustically robust egg-crate geometry for the tower. The Tower bears on a transfer structure at second floor level to allow minimal vertical structure below so that the public spaces can flow through the whole ground floor.

BUR JUMAN CENTRE

Location: Dubai, United Arab Emirates
Date: 2005
Architect: Kohn Pedersen Fox Associates
Client: Bur Juman Centre LLC
Buro Happold services: Structural engineering, building services, ground engineering, infrastructure and transport engineering, fire engineering, CoSA (Computation and Simulation Analysis)

The refurbishment of this luxury retail complex in the centre of Dubai comprises new retail malls as well as four levels of basement car parking, in all some 50,000 square metres. Refurbishment of existing malls – 45,000 square metres – includes a multiplex cinema, a 23-storey office tower and conference facility, and two top-quality apartment towers of 15 and 19 storeys. The new shopping centre features a number of leading international outlets and a health club, children's crèche, as well as internal landscaping and extensive rooftop gardens.

The glazed and louvered roof structures allow in controlled levels of daylighting to minimise heat gain.

Roof structures over shopping mall

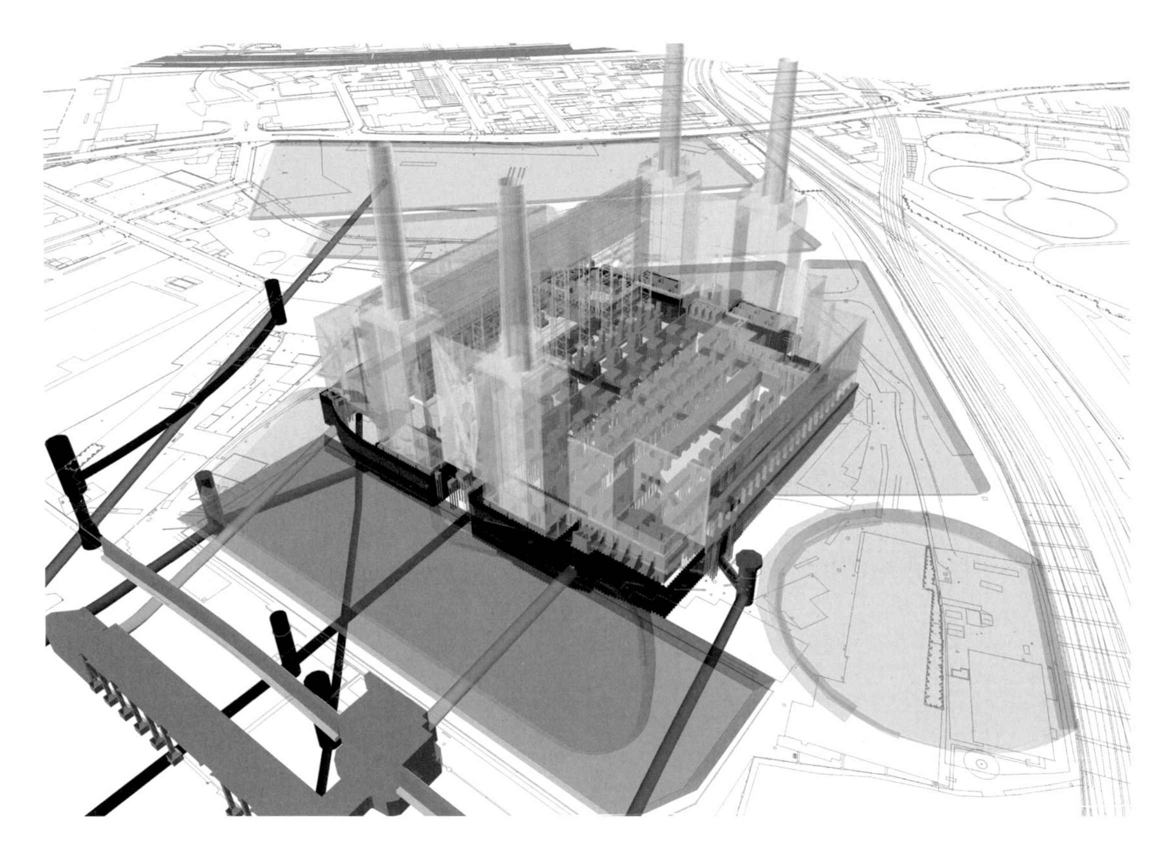

Computer model of new masterplan for Battersea Power Station

External view of the existing Battersea Power Station building

BATTERSEA POWER STATION MASTERPLAN

Location: London, UK
Date: 2020
Architect: Rafael Viñoly Architects
Client: REO (Real Estate Opportunities) Ltd
Buro Happold services: Structural engineering

Battersea Power Station was decommissioned back in 1982. Subsequent plans for an indoor theme park floundered but the site has now been given a new lease of life as a mixed-use development which will form one of the UK's largest brownfield / urban regeneration projects.
The masterplan will provide approximately 750,000 square metres of residential, office and retail space. There will be a 2.4 hectare public park, a riverside walk and an urban square.
The intention is to create a sustainable scheme in keeping with the importance of the site and the heritage building, which is both commercially viable and deliverable.

The development will eventually include both original and new buildings, including a proposed landmark commercial construction next to the power station that is intended to be the most sustainable building in the UK.

Another fragmented old building that is to be returned to use is Battersea Power Station on the banks of the Thames. Built in two phases in the 1930s and 1950s, the building was once thought to be the largest brick structure in Europe, but Giles Gilbert Scott's masterpiece of industrial architecture was decommissioned as a power station in the 1970s. In the 1980s, the Grade 1 listed building was partly destroyed as part of the first stage of an aborted crass scheme to turn the site into an amusement park. Now, a new plan by REO (Real Estate Opportunities) Ltd will incorporate hotel, residential and retail accommodation, while key historic spaces will remain open to the public. New buildings will be added to the site, but the remains of the original masterpiece will be retained and the old massing will be restored.

So the familiar shapes of the turbine halls and boiler houses will be saved, with the massive white concrete chimneys at each corner of the turbine hall (hence the local nickname 'the upturned table'). To optimise the costs of the large development, as much existing structure as possible is to be re-used, and Buro Happold undertook a structural survey. Though much of the building complex had been demolished, foundations still exist, but records of them were lost. So their form and location were determined by geophysical investigation using such techniques as electromagnetic and ground probing radar, sonic echo and seismic survey to determine the structural integrity and depth of piles. A three-dimensional picture of the underground structure and its structural properties was built up, waiting for the next phase of development.

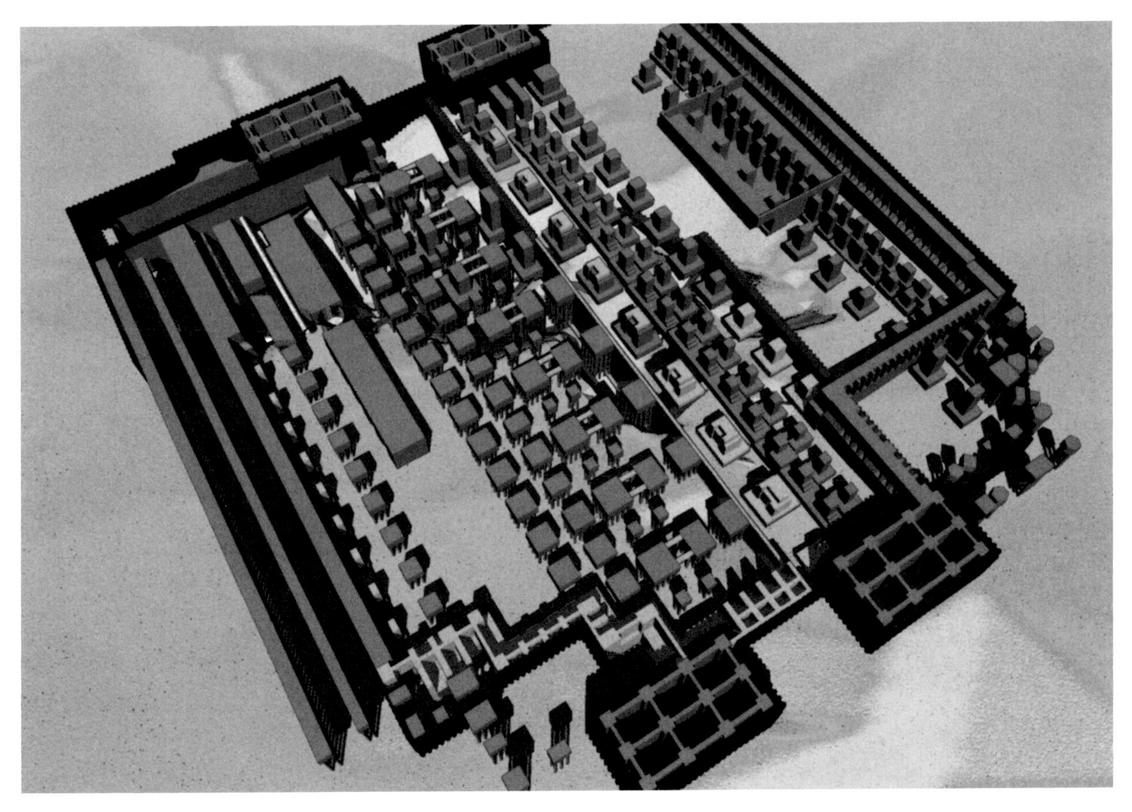

Rendering of existing foundations that will be re-used within the new scheme

Modern cities are under pressure to increase their density because otherwise their footprint will spread infinitely as their populations increase. Further, properly designed dense cities have lower transport needs than scattered ones and they can be more welcoming to pedestrians. One of the commonest ways of increasing urban density is to build high, as places as different as Manhattan, Hong Kong and downtown Sydney demonstrate. In Abu Dhabi, Kohn Pederson Fox working with Buro Happold responded to the seaside site with a 37-storey headquarters office tower for ADIA that consists of two curving wings connected by a vertical courtyard atrium. One wing aligns with the tight urban grid (the city is the only one in the United Arab Emirates that has a very strong plan). The other wing peels back to make views of the sea and the direction of Mecca available right from the middle of the building. The twin glass façades are designed to reduce heat loss and gain to the interior by allowing air extract behind the third layer of glass.

Aerial view of ADIA and adjacent site of the landmark tower

ABU DHABI INVESTMENT AUTHORITY (ADIA)

Location: Abu Dhabi, United Arab Emirates
Date: 2006
Architect: Kohn Pedersen Fox Associates
Client: Abu Dhabi Investment Authority
Buro Happold services: Structural engineering, building services, cost consultancy, project management, ground engineering, civil engineering, fire engineering

The stunning design, incorporating a stylish twin-wall glass façade and dual curved office wings, complete with central atrium, was immediately hailed as an engineering and architectural success. After originally construct a 33-floor tower, a further four floors were added to the office and the building services capabilities increased. Careful attention was given to modifying the external landscape to accommodate the additional people and car parking – ADIA can now cater for a population of 2,000 with 800 parking spaces.

The 200-metre-high structure is in a premier city centre location and includes restaurant, gym and swimming pool facilities for employees. Infrastructural innovations include a link from the car park to the underground train system. Notable features include a 180-metre-high atrium wall and a 30-metre cantilevered steel grillage canopy.

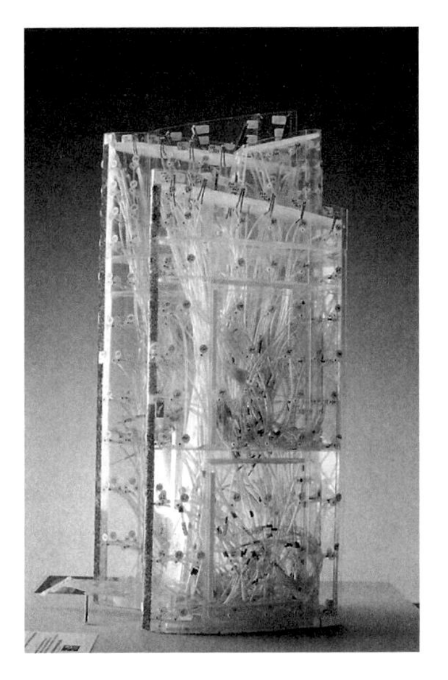

Wind tunnel model

The office façades nearing completion and the sloping façade over the atria that lie in between.

This view of the triple glazed façades with vertical modulating fins shows the process of enclosing the reinforced concrete floor plates.

The Singapore National Library is as responsive a solution to programmatic and climatic imperatives as the Abu Dhabi tower; so, of course, results are very different. Ken Yeang, principal in charge of T.R. Hamzah & Yeang's architectural team, has long been an expert in ecologically aware architecture for the tropics, and the project, on which Buro Happold did the services and structural engineering, incorporates lessons learned by both practices. The brief called for two kinds of accommodation: a fine library with the usual storage, study and curatorial services, and a more mixed (and perhaps noisy) group of functions, including exhibitions, multi-media displays and lectures, conferences and concerts.

The designers decided to plan the building on its very tight inner urban site as a couple of blocks separated by a day-lit, semi-enclosed internal street and linked by bridges over it. Library collections and their ancillary activities are in the larger block. Across the pedestrian street, the other elements of the complex are housed in a thinner curved slab. A piazza for public events is formed in the middle of the street, a communal space intended to become the main focus of the site. It is surrounded by cafés, a library shop and other retail spaces.

External view

NATIONAL LIBRARY OF SINGAPORE

Location: Singapore
Date: 2007
Architect: T. R. Hamzah & Yeang
Client: National Library Board, Singapore
Buro Happold roles: Building services, structural engineering

The new state-of-the-art National Library in Singapore provides an open, welcoming learning environment and serves as an icon for the region, reflecting the city's multi-cultural heritage. Located in the central business district of the city, the library consists of two buildings that are separated at ground level by internal streets and joined by bridges at the upper levels. The division of the library into two reflects the need to present it as a unique and innovative cultural facility, while the naturally ventilated civic plaza provides a public meeting place and a space for outdoor events.

One of the naturally ventilated civic plazas dedicated to outdoor activities and green space

View of glazed façade of one of the internal streets that connect the two buildings

Elevations of the two blocks are given expression by sunshading systems, external louvre blades, some as much as 6 metres deep, creating a tropical aesthetic that is quite new, yet related to the immemorial shading devices of local vernacular architecture. Throughout the building, three strategies of environmental control are employed for different functions. First, passive energy design, used as much as possible, optimises daylighting, makes maximum use of orientation and natural ventilation and of planting with native species. Second, an active strategy incorporates full air conditioning and artificial light; this approach is mainly reserved for the library stacks and study areas. The third strategy mixes the other two in systems that for instance use fans to supplement natural ventilation in lobbies, foyers and the piazza. Due to these strategies, the National Library has a lower environmental impact than a conventional Singapore office building of the same size, even though it has in some places got much higher environmental control requirements.

The National Library is a state monument and, though it will undoubtedly be changed in future, it will largely retain its fundamental configuration for many years. Yet because of intense pressures for alteration caused by rapid technological change and ever increasing urban populations, many modern urban buildings will have to be capable of radical transformation, movement even. Shigeru Ban's Nomadic Museum is a travelling structure to house a photographic exhibition by artist Gregory Colbert. Ban made the walls of used freight containers, and the roof is supported on his well-tried cylindrical cardboard columns. The museum was first erected on top of Manhattan's Pier 54 on the Hudson River at 13th Street.

Urban model of the corals

Model view

CORALS OF SHARJAH

Location: Sharjah, United Arab Emirates
Date: 2006
Architect: Henning Larsens Tegnestue
Client: Prince Khalid S.M. Al Saud, Kingdom of Saudi Arabia; Ali Al Murbati, Sharjah
Buro Happold services: Consulting engineer

The Murjan (Coral) of Sharjah is a unique tower complex consisting of five independent towers rising from a public peninsula south-west of Sharjah. The complex, proclaimed to be the tallest building in the world, has a total floorage area of 4.2 million square metres and a height of approximately 1 kilometre. Placed on the borderline between sea and desert, the coral towers form an organic whole reflecting the very nature of towers in the Islamic tradition.

In order to create such high towers with a distinctive appearance, this concept design for the external mega-structure was based on the natural forms of coral. This primary load-bearing shell to the outside façade of the tower is an interpretation of traditional patterns at various scales.

The outer walls were formed of 148 used shipping containers arranged in a chequerboard pattern; 37 carried the exhibition from place to place and the rest were hired on the spot. Very simple triangular roof trusses were made of 305 millimetre diameter cardboard tubes bearing onto paired 736 millimetre diameter cardboard columns. Tent-like fabric filled the gaps between containers and formed the roof covering. Care had to be taken to ensure effective structural continuity between the paper structure and the containers. The main problem was resisting lateral loads on the walls, lightweight because the containers were empty.

In the end, the museum was a triumph, in which carefully orchestrated artificial lighting by Alessandro Arena converted the interior into a basilica-like space, with a brightly-lit nave and two shadowy aisles. The photographs were suspended between columns, and a diaphanous curtain made of a million Sri Lankan tea bags (recyclable materials again) modified the space and added to its almost religious intensity. The building's basic thesis was successfully demonstrated when it was moved to California in a reconfigured plan in 2006.

NOMADIC MUSEUM

Location: New York, NY, USA
Date: 2005
Architect: Shigeru Ban
Client: Bianimale Foundation
Buro Happold services: Structural engineering

Originally located on New York's historic Pier 54, where the Titanic was to have docked, the Nomadic Museum was designed to be a temporary structure that can be dismantled and reassembled in a series of cities around the world. The frame is made up of 148 empty shipping containers, stacked in a self-supporting grid, while waterproof paper tubing is utilised in the columns and roof trusses. The pitched, fabric-covered roof consists of simple steel girders spanning from the outer containers to the inner support columns. As an integral part of the aesthetic experience, the building successfully frames a poetic context for viewing 200 large-scale artworks by photographer Gregory Colbert. With no natural light inside, the photographs appear to float in a mystical space.

Building in its first location on a pier on the Hudson River, New York. Today it is in a new form situated in California.

View between the fair-faced precast concrete stellae showing narrow, 0.95-metre-wide pathways, providing enough space for only one person to pass along at a time

MEMORIAL TO THE MURDERED JEWS OF EUROPE

Location: Berlin, Germany
Completion: 2005
Architect: Eisenman Architects
Client: Foundation for the Memorial to the Murdered Jews of Europe
Buro Happold services: Structural engineering, building services, site infrastructure

The site for the memorial is located in central Berlin, between the Brandenburg Gate and Potsdamer Platz. Following Peter Eisenman's architectural design, a field of 2,711 concrete stelae and an underground information centre were constructed on the 19,000-square-metre site. The stelae measure 2.38 x 0.95 metre and vary in height up to a maximum of 5 metres. Buro Happold was responsible for developing a meticulous and innovative methodology for constructing the stelae in order to meet the architect's demands and to ensure durability over time.

At the memorial site, the width measurements between the rows of stelae are restricted, leaving a pathway narrow enough for only one person to pass through at a time.

Image on previous page: Aerial view of the Memorial to the Murdered Jews of Europe

One of the most exciting developments in the three-dimensional arts over the last 50 years has been the rapid evolution of landscape – or land art as it was called in the USA, where the movement started in the 1960s. Suddenly, an art that had been immensely important in the 18th and 19th centuries, but which had apparently fallen into decorative slumber in the first part of the 20th century, became inventive and potentially moving. Its practitioners use the traditional materials of the landscape architect – earth, plants, rocks, water and so on – to create works that can have the sort of emotional power as the great landscape gardens of the 18th century. While landscape design, almost by definition, appears to be an innocent pursuit, manipulation of the natural world has always involved large amounts of energy. The gardens and waterworks of Hadrian's Villa at Tivoli could not have been made without very thoughtful engineers and brigades of slaves; Lancelot (Capability) Brown's seemingly natural 18th century English landscapes needed platoons of navvies to move hills and create lakes.

View of the site boundary showing the undulating field of stelae, each one measuring 2.38 × 0.95 metre. Some reach a height of 5 metres as the ground slopes away towards the centre of the site.

Modern machinery and technology have made large numbers of labourers unnecessary, but engineering skills remain vital, as designs become both more subtle and increasingly ambitious. One of the largest and best-known recent examples of landscape transformation is New York architect Peter Eisenman's Memorial to the Murdered Jews of Europe, near the Reichstag and the Brandenburg Gate in Berlin. The task was prodigiously difficult. Eisenman said that he wanted the place 'to speak without speaking', and the monument consists of 2,711 mute concrete pillars (stelae) arranged on a rigorous grid of one-person-wide (0.95 metre) paths that divide the 0.95-metre-broad stelae that have a long plan dimension of 2.38 metre.

The ground sinks gradually towards the middle of the site, but the stelae vary in height, from being practically flush with the earth at the edge of the Field of Memory to 5 metres high in the middle of the monument – an overwhelming place where you are dwarfed by the stelae. These are set at slight but different angles to the vertical, giving the unnerving hint that you are surrounded by a faceless crowd clad in Prussian field-grey uniforms or, perhaps, concentration camp clothing: Eisenman is trying to reveal what he calls 'the enormity of the banal'.

Aerial view of the six buildings and street that form the overall site

BRITISH AIRWAYS HEADQUARTERS

Location: Waterside, Harmondsworth, UK
Date: 1998
Architect: Niels Torp with RHWL Architects
Client: British Airways
Buro Happold services: Structural engineering, ground engineering, geo-environmental engineering, roads and parking, project management

Waterside is British Airways' large corporate headquarters, built on 112 hectares of land formerly used as a domestic waste tip at Harmondsworth, near Heathrow Airport. The development comprises six horseshoe-shaped buildings which all back onto a quadruple height street. Waterside is based on a village concept, with individual neighbourhoods situated along a central internal street.

The six individual buildings are all four-storey and have between 7,000–9,000 square metres of gross internal area. They are arranged on either side of the landscaped atrium street. This is a linear plaza (175 metres long and 12 metres wide) paved in granite slabs, granite setts and cobblestones. It forms the building's heart. The buildings and street are linked by a glazed roof which oversails the street.

As structural and services engineers, Buro Happold had to organise the gentle curves of the ground form which are entirely artificial – left untouched, the site would be a flat, sandy waste. The monument's large paved surface needs a complicated but inconspicuous drainage system. Run-off water is taken to an underground reservoir for irrigating trees planted among the low stelae on the west side of the site, just across the road from the Tiergarten. An underground information centre provides the words and images that are so remarkably absent from the Field of Memory. Its roof is a reflection of the grid of the stelae and the artificial undulation of the field overhead.

Another tricky engineering problem was the manufacture and erection of the stelae. They are almost perfectly formed monolithic concrete boxes which lack a bottom surface, and are supported on strip foundations. They had to be shimmed to the exact height and angle (up to two degrees from the vertical) specified by the architects for each one. Cast in self-compacting concrete, the stelae had to be specially cured before being delivered to the site; weathering trials lasted a year. A waterproof coating is covered by an anti-graffiti layer – needed because of the activities of local youths, who can approach the Field of Memory at any time, for the monument has no boundary fence. Instead, it emerges gently, yet powerfully out of the very soil that nurtured the horrors that necessitated its existence.

Glazing to the landscaped street at British Airways Headquarters

Detail of the façade between buildings which includes controllable openings for natural ventilation

British Airways Headquarters at Harmondsworth near Heathrow by the Norwegian architect Niels Torp, with Buro Happold as structural and geotechnical engineers, is another example of large-scale earth moulding, but of course it is completely different from the Berlin memorial in location, purpose and expression. The site, next to the airport, was contaminated by earlier industrial uses; it had four closed and two open landfill sites for domestic rubbish. Further, the site is bounded on two sides by rivers and the water table is very high.

The offices (in six four-storey blocks arranged round courts) accommodate some 3,000 people. A well-lit internal street with shops and cafés links the four blocks and provides a common social area. Because the site is not well served by public transport, and it has a population as big as that of a village, extensive car parking was necessary, most of which is underground below the buildings and parts of the landscape. Some 3 million cubic metres of waste and clean materials had to be moved to create 120 hectares of reconstructed landscape, a large part of which is a public park.

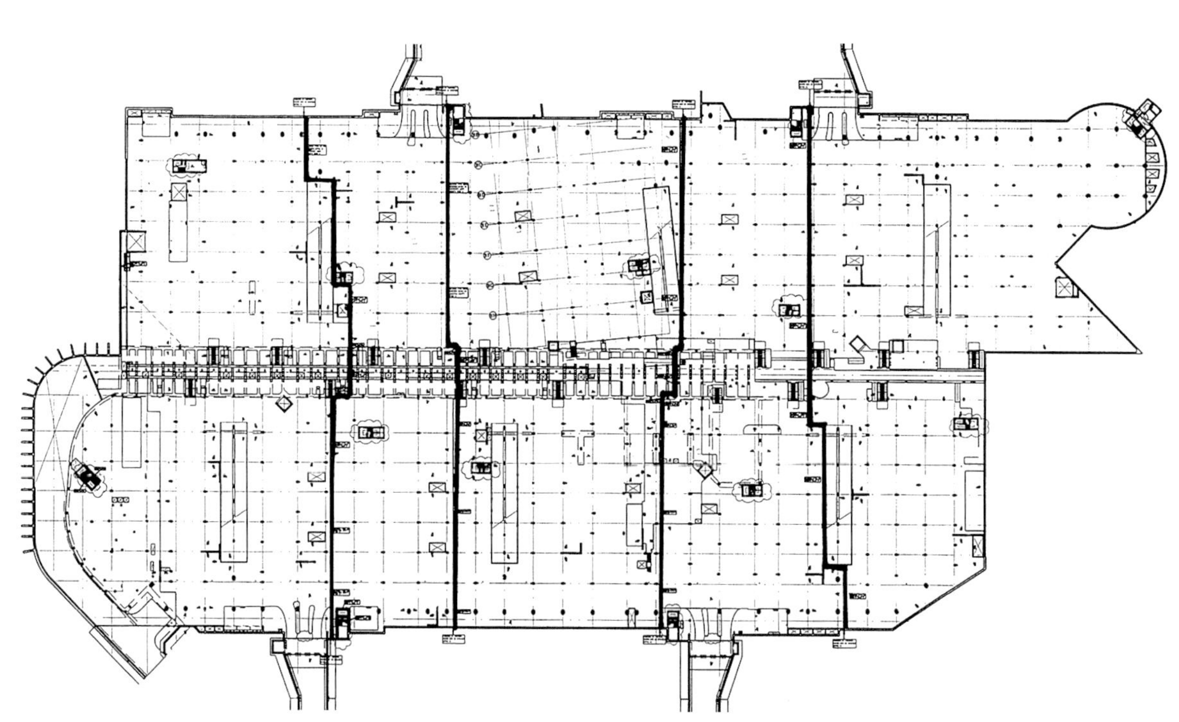

Plan of the basement car park showing the entrance at left and central service-way

To prevent the car parks from being flooded, and to avoid penetration of gas and leach-liquid from neighbouring landfills, a clay berm seal was put round the building site and toed into the underlying London clay. Clay was also used to cap off the landfill areas on the adjacent parkland site. A perimeter drainage system ensures that basements are dry, and that the buildings have no tendency to float on underground water. The offices themselves are long in plan, and precautions had to be taken to minimise movement of their concrete frames as the varied ground conditions gradually change. Among tactics to avoid stresses above ground are sliding joints at the heads of basement columns that allow horizontal movement. None of this is visible from the surface, where a luxuriant and varied landscape enhances the buildings' views and courts – and creates a generous addition to the village of Hounslow, sandwiched between the industrial zone and the great airport.

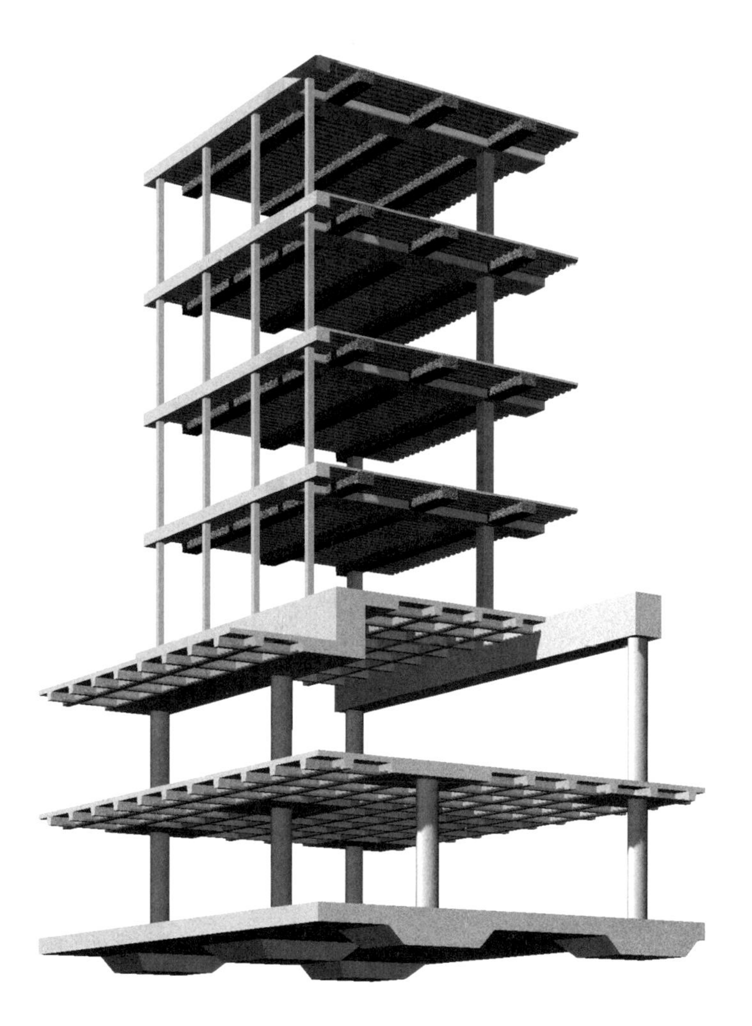

Isometric of structural systems at British Airways Headquarters, showing the composite metal decking precast and in-situ beam strips to four storey offices, above basement car parking waffle slabs

At the Centre for Mathematical Sciences at the University of Cambridge, Edward Cullinan Architects, with Buro Happold as consulting engineers, decided to make the landscape climb up over the building. As at the British Airways Headquarters, office accommodation is in individual pavilions (square rather than stretched out) that are connected by common spaces. To start with, a basin was dug out of the site so that the mass of the new building would not dwarf surrounding villas. The major space in the maths centre's central core is a generous, if rather cavernous top-lit cafeteria organised to encourage informal meetings and discussion.

Externally, the cafeteria's roof is covered in grass to provide pleasant views from the upper floors and to be a place of relaxation for the building users on fine days (in summer, its gentle green curve is liberally strewn with sun-bathing students). Structurally, the shape of the roof is achieved with a curved slab in concrete (used because it is relatively easy to waterproof) supported by concrete-buttressed concrete and precurved steel arches. These have high-grade cast-iron pins at their springing points, partly so that their loads can be analysed accurately and partly to ensure clarity in the arches' form, allowing the cafeteria to seem like the interior of the elegant rib-cage of the great giant Mathematics.

Cafeteria building: concrete-cased steel arches of diminishing span rest on in-situ concrete buttresses, showing clerestory lighting and natural ventilation

Turf roof in foreground to cafeteria with heavy weight concrete and stone buildings for the offices

CENTRE FOR MATHEMATICAL SCIENCES

Location: Cambridge, UK
Date: 2003
Architect: Edward Cullinan Architects
Client: University of Cambridge
Main contractor: Sir Robert McAlpine Ltd
Buro Happold services: Structural engineering, civil engineering, geotechnical engineering

The site comprises a series of individual pavilions placed round a shared central space. The building has been designed to take advantage of excavated space to keep the overall height of the centre to an acceptably low level. The shared central core will function as a connecting space between the pavilions, and will include entrance court, lobby, common room and dining room at ground level. Most of the shared accommodation is placed in a lower ground level. Lecture rooms, seminar rooms and laboratory are placed on the lower level, below ground.

Drawing of the supporting pin to the cafeteria arches

From earliest times, humanity has struggled to abate and even tame the huge energies of the sea. A large part of the history of civil and structural engineering is concerned with maritime structures, a tradition that flourishes today, in, for instance, the Chatham Maritime project. Chatham, on the Medway, a tributary of the Thames, was one of the great depots of the British Royal Navy for three centuries, and many of the Navy's most famous ships were built there. Luckily, when the dockyard was closed in the 1980s, the awesome sheds in which the vessels were built were still standing, and they, with their ancillary buildings, will be turned into a naval museum. Much of the rest of the site is being redeveloped with mixed uses.

The most dramatic new elements are the two housing towers on the quays that dramatically mark the mouth of the harbour. Designed by architects Wilkinson Eyre with Buro Happold as structural engineers, the 15- and 19-storey towers are the most conspicuous parts of a large mixed development that includes residential, commercial and leisure accommodation. Victorian dock walls constrict the developable area available round each tower, so loads have to be transmitted as directly downwards as possible. Both towers have a basement box for car parking and plant, so foundations are almost completely below water level; 58 bored piles below each box carry loads through layers of gravel down to the underlying chalk 40 metres below.

Visualisation of the completed development from the triangular footpath

CHATHAM MARITIME

Location: Chatham, UK
Date: 2010
Architect: Wilkinson Eyre Architects
Client: ING Real Estate
Buro Happold services: Structural engineering, transport assessment, building services, ground engineering, fire engineering, acoustics, sustainability

Chatham Maritime is a major mixed-use development that occupies a large site close to the Chatham dockyards on the Medway River. Two distinctive glazed residential towers will stand out as landmarks, almost like lighthouses, positioned on the northernmost tips of the peninsula's two historic quays. With curvilinear triangular footprints, the 15- and 19-storey-high towers, oriented east and west, will capture stunning views of the estuary and river.

They stand above a further development of two low-rise industrial-style buildings, one accommodating more apartments, the other designated for commercial and leisure use. Buro Happold's ground engineering team had to develop an innovative piling scheme to support the towers without impacting on the dock walls and taking into account the site's ground conditions. Structurally, windloading is a clear problem for the towers on their exposed site.

Visualisation of the
completed development

In plan, each tower is a rough isosceles triangle with smoothly rounded corners (the latter partly to reduce wind loads). Tower stiffness is imparted by a vertical circulation core at the rear apex of the plan. Further resistance to the lateral and torsional forces that can be expected at such an exposed nautical site is given by concrete shear walls, which transfer the centre of stiffness to the approximate geometrical centre of the plan triangle. Such moves were not enough to withstand some of the wind loads that can be expected, so eight raking piles are added to the foundations below each tower. They bear into the gravel layers by passing (in some cases) under the dock walls, a delicate move requiring much care of the old structures.

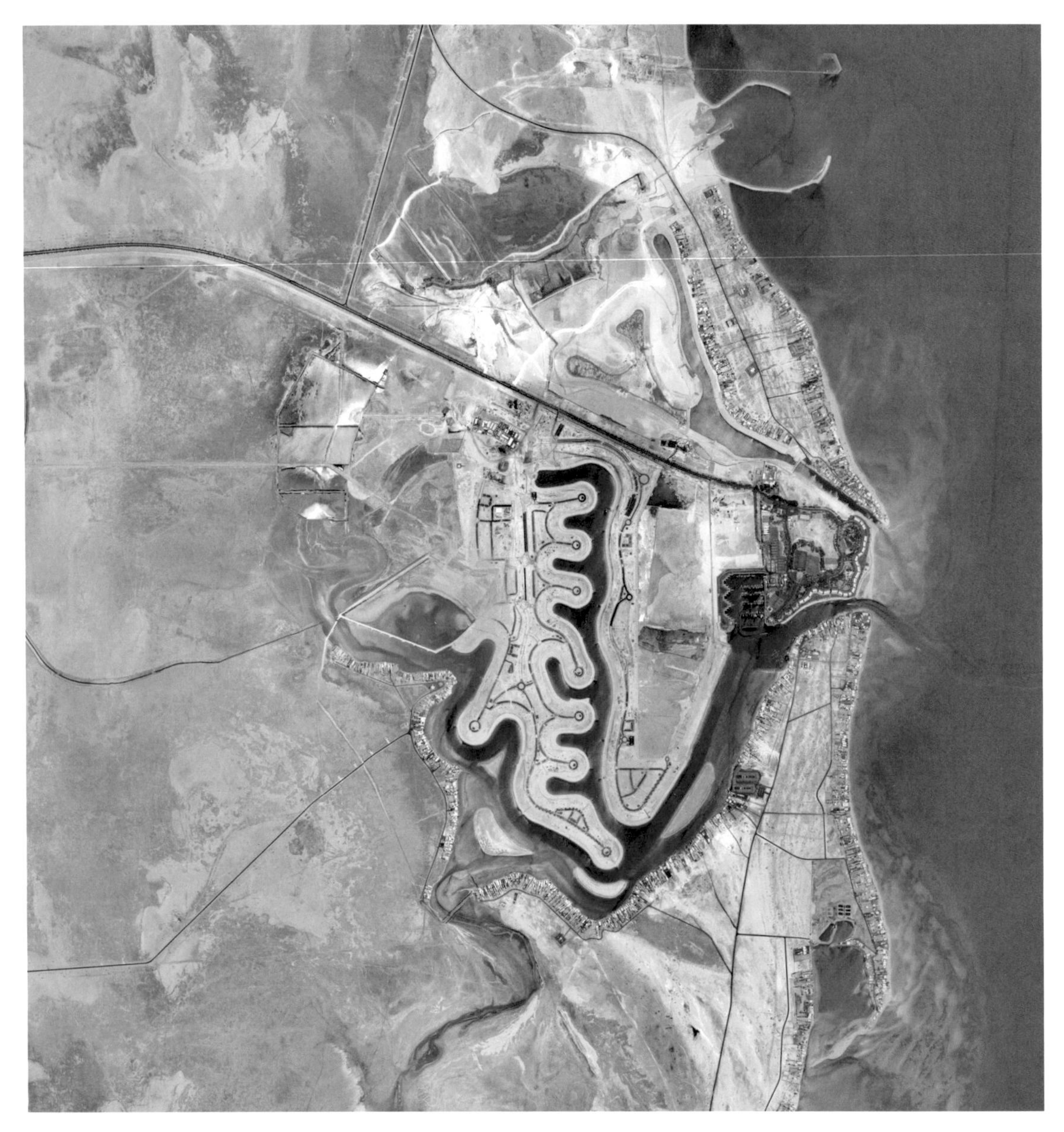

Aerial view of the 6,400-hectare site

At Chatham, the site was determined by the historic dockyard: at the Al Khiran Pearl City in Kuwait a new site had to be created. Now that Gulf cities have become popular as tourist and business centres, there are huge demands for coastal locations, but there is simply not enough coastline to go round. So schemes have been developed to extend the shore's length. At Al Khiran, Buro Happold, working with Gulf Consult and Coastal Science Associates, were responsible for the overall plan, marine and highway engineering. The consultants developed a plan that strongly resembles a vastly magnified portion of a mammal's lung, with convoluted organic geometry designed to maximise exposure to the pervasive medium: air for lungs, the sea for Kuwait. The first phase has already more than doubled the area of waterways within the Al Khiran estuary. Indigenous species of marine flora and fauna are beginning to flourish: their progress is being monitored as part of the project.

Ecological concerns dominate another of Buro Happold's Middle Eastern projects. Riyadh, the capital of Saudi Arabia, was founded in the desert on a natural watercourse, the 120-kilometre-long Wadi Hanifah which, like all Arabian wadis, is normally dry, but subject to periodic flash flooding every five or ten years. As the city grew, the lower reaches of the wadi became a flowing river formed from sewage effluent and rising groundwater. The urban population is expected to rise from 5 million in 2004 to over 10 million in 2021, so conditions were predicted to get much worse.

AL KHIRAN PEARL CITY

Location: Al Khiran, Kuwait
Date: 2030
Client: La'ala Real Estate Company
Buro Happold services: Masterplanning, coastal modelling, earthworks and coastal engineering, transportation planning, highway engineering, infrastructure design, bridge engineering, structural engineering, quantity surveying, construction management, site supervision

The site of the Al Khiran Pearl City development is on the Arabian Gulf coastline, some 85 kilometres south of Kuwait City. There are two natural tidal creeks at the site, and half the 6,400-hectare area is salt marshland. The planned duration of development of the city is expected to be 25 years, with development in some ten phases.

The final population of the city is anticipated to be 100,000. The primary objective is to develop Al Khiran Pearl City as a unique and attractive waterside development that encompasses local cultural traditions, tempers the harsh environment with good quality construction, landscaping and recreation facilities and sets a high standard for coastal development in the Middle East.

The reinforced-concrete water towers will provide the potable water supply to the new development.

Rendering of the finished development adjacent to the new marina

Date palm groves at the base of Wadi Hanifah

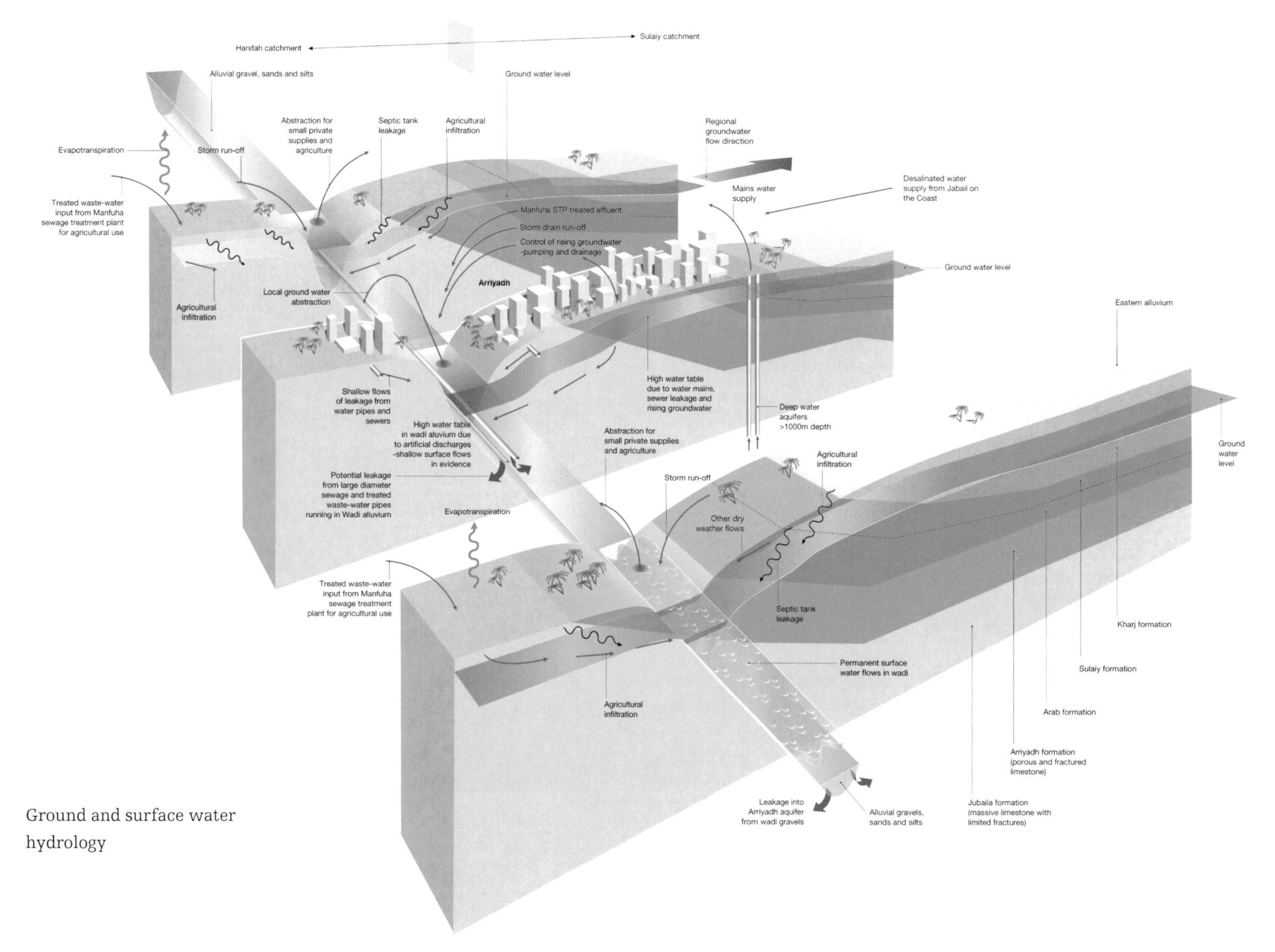

Ground and surface water hydrology

Buro Happold worked with architects Moriyama & Teshima to create a ten-year plan that will improve water quality, restore the landscape and reduce reliance on extremely expensive (in both monetary and energy terms) desalinated water produced on the coast. Biomediation techniques will be used to improve the quality of water in the flowing part of the wadi, so 250,000 cubic metres of water will be available daily for irrigation and other non-potable uses. By 2021, the amount of such re-usable water should rise to 1 million cubic metres. The wadi flood channel will be returned to its natural function (and where possible its natural form). Landscapes will be restored with native species, and infrastructure will be integrated and upgraded to resist floods. Water produced by the biomediation process will radically reduce the need for the desalinated product. Natural treatment is reckoned to cost only a third of traditional water cleansing technology, and it will radically reduce electricity demands for distilling sea water. The scheme aims to integrate the city, its people and the wadi, which was the original reason for the settlement, and holds out hope for more biologically-based sustainable environmental control technologies in future, particularly in Arabia.

WADI HANIFAH

Location: Riyadh, Saudi Arabia
Date: 2007 (phase 1)
Masterplan: Buro Happold with Moriyama & Teshima Architects
Client: Arriyadh Development Authority
Buro Happold services: Masterplanning, river and hydraulic engineering, civil engineering, transportation planning, ground engineering, infrastructure, quantity surveying, site management, bridge engineering

The masterplan's focus is on restoring the wadi's natural beauty, which has been spoiled by decades of unrestricted dumping and development, but it is also intended to harness and rehabilitate water. The plan provides the potential for recycling 1 million cubic metres of water per day by 2021 to meet a third of the city's water demand. The project is complex, due to the technical challenges facing the many individual engineering disciplines and the co-ordination of these elements into a restoration project.

It involves three major components: an appraisal of current conditions, a comprehensive development plan and an implementation programme. This is an extensive development that will take ten years to complete. The first phase, to remove 1.25 million cubic metres of rubble from the valley, is in progress.

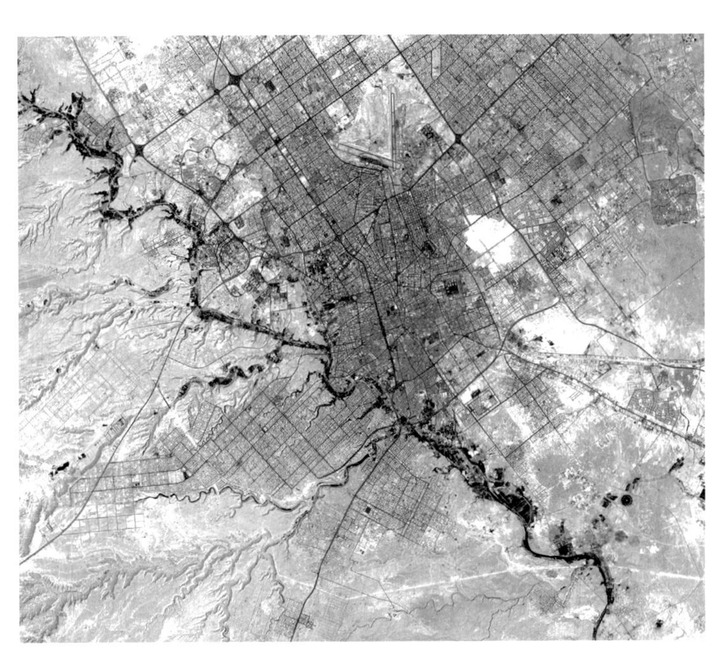

Aerial view of the 40-kilometre wadi

View of Hayle Harbour coastline

Aerial view of the redevelopment site

HAYLE HARBOUR REDEVELOPMENT

Location: Penwith, Cornwall, UK
Date: 2012
Architect: FSP Architects and Planners
Client: ING Real Estate Developments (Hayle Harbour) UK Ltd
Buro Happold services: Rivers and coastal engineering, flood risk assessment, site infrastructure, environmental impact assessment, bridge engineering, planning supervision

ING acquired the site in 2004 with the objective of regenerating the harbour to bring it back into effective and efficient use. As well as restoration of the harbour infrastructure, the project will provide a permanent floating harbour with navigation lock, a tidal flood defence to the development and the town of Hayle, a marina, a fishermen's harbour, and a mixed land-based development including 700 residential units, shops, hotels, commercial and business facilities.

The development will be underpinned by improvements to the existing infrastructure, including the provision of roads, parking, bridges and utilities. The current phase includes the development of a masterplan for the regeneration scheme. Thereafter, the project will move to detailed design, with construction due to take place over a number of years once planning permission has been obtained.

Internal view of 'the Hub', a public mall which is housed inside the Millennium Point development

MILLENNIUM POINT

Location: Digbeth, Birmingham, UK
Date: 2001
Architect: Nicholas Grimshaw & Partners
Client: Millennium Point Trust
Buro Happold services: Structural engineering, transport engineering, fire engineering

The development covers a site the size of six international soccer pitches, and is the catalyst for the regeneration of the east side of the city centre. Millennium Point provides a home to Thinktank, an interactive exploration taking the visitor into the world of sciences, technology, medicine, nature and our everyday lives. The enormous complex also houses 'the Hub', a public mall, the University of the First Age and Young People's Parliament, the region's first IMAX Theatre and the Technology Innovation Centre.

The site is within five minutes walk of Birmingham city centre, adjacent to the campus of Aston University and Aston Science Park. From the start of the project, an essential aspect of the brief to the design team has been to create a high-quality modern building that expresses the engineering heritage of the region.

View of the elegant
S-shaped crossing

Bronze posts line the
crossing, acting as
a handrail

SACKLER CROSSING,
ROYAL BOTANIC GARDENS

Location: Kew Gardens, London, UK
Completion: 2006
Architect: John Pawson
Client: Royal Botanic Gardens
Buro Happold services: Bridge engineering, infrastructure, ground engineering

The Sackler Crossing is situated at the west of Kew Gardens and provides access across a small lake to a previously underused area. The bridge's design, is intended to blend in with its setting. The bridge has a deck of black granite sleepers supported at a minimum height above the surface of the water. It curves round an island in the lake in a gentle S-shape and is lined by almost 1,000 bronze posts which act as balustrades. Each of these bronze fins are smoothed and contoured to fit in the pedestrian's hand, instead of a rail across the 70-metre bridge. One of the key visual aspects of the bridge is the 564 black granite sleepers, which measure 120 × 120 millimetres by 3 metres and weigh 130 kilograms each.

To accommodate the curve of the bridge, the gaps between them taper across the width of the deck. These are precisely bolted onto the steel structural base, which rests on nine 457 millimetre diameter driven steel piles. The 990 uprights are made of aluminium bronze, which has the appearance of gold and has excellent corrosion and abrasion resistance. Each upright was cast and then flattened and finished to form a rounded, ergonomically contoured top. The uprights each have a gap of 10 centimetres between them which, depending on the angle of viewing, appears and then disappears.

Bridge heading to the Clarence Dock development

CLARENCE DOCK REDEVELOPMENT

Location: Leeds, UK
Date: 2007
Architect: Carey Jones Architects
Client: Berkeley Clarence Dock Company
Buro Happold services: Building services, structural engineering, ground engineering

This mixed-use development lies on a 6-hectare brownfield site next to the River Aire and close to Leeds city centre, and will provide 1,000 luxury apartments, 10,000 square metres of commercial office space and 50,000 square metres of leisure, exhibition and retail amenities with associated parking facilities. Among the buildings within the developments are a hotel, casino, residential and office towers. Buro Happold is appointed to work on a variety of residential and retail towers, one of which will be 20 storeys high, a six-storey office building, casino and retail units. In addition, the firm is designing a footbridge across the River Aire and two moveable bascule bridges in the dock area nearby.

Residential apartments at Clarence Dock

Section through basement and superstructure

Base of finished central atrium showing principal supporting structure

PALACE OF PEACE AND RECONCILIATION

Location: Astana, Kazakhstan
Date: 2006
Architect: Foster + Partners
Client: Sembol Construction
Buro Happold services: Structural engineering, fire engineering, risk assessment

In September 2003, Kazakhstan, the largest of the former Soviet Republics, hosted the inaugural Congress of Leaders of World and Traditional Religions in the new capital, Astana. Spurred by its success, the President of Kazakhstan decided to make it a triennial event. Foster + Partners was commissioned to design a permanent venue for the congress: the Palace of Peace and Reconciliation. The construction schedule was extraordinarily rapid: the Palace was to be completed in time for the meeting of the second congress in 2006.

This led the design team to develop a structural solution that utilised prefabricated components, which could be manufactured off-site during the winter months and erected during the summer. The entire process, from briefing to completion, took just 21 months.

Image on previous page: Internal view of Palace of Peace and Reconciliation

What of the future? Buro Happold is now a large firm with 25 offices in four continents working on a wide and diverse range of projects. Naturally, exploration continues into new ways of resolving relatively conventional problems, as well as into uses for new materials and into innovative approaches to structure and environment. In the Palace of Peace and Reconciliation in Astana, Kazakhstan, the architects, Foster + Partners, and Buro Happold (working as co-consultants with Arce of Turkey) were met with what at first must have seemed to be a very daunting set of predicaments. In 2003, president Nazarbayev of Kazakhstan held a conference of leaders of the world's faiths which he judged to be such a success that he decided to repeat it triennially. Foster's design was chosen for the meeting place, but the programme was so tight that the whole massive building had to be completed in 21 months from the architects' initial briefing.

Isometric of the grandstand structure

ASCOT RACECOURSE

Location: Windsor, UK
Date: 2006
Architect: HOK SVE
Client: Ascot Racecourse
Buro Happold services: Building services, structural engineering, ground engineering, lightweight and long-span structures, site infrastructure, contaminated land and remediation assessment, fire engineering, people flow consultancy, lighting design

The Ascot Authority's aspiration was to redevelop Ascot as 'the finest racecourse in the world', providing spectators and visitors with world-class racing and viewing facilities. The project entailed a thorough reorganisation and rebuilding of the facilities. The project comprised two main components: the enlargement of the current site and the construction of a new grandstand.

The existing grandstand has been replaced by a new and larger one, which provides greatly improved facilities for race-goers. This spectacular structure comprises viewing areas with terraces and private boxes, plus relaxation areas with dining rooms, bars and lounges. An elegant long-span roof cantilevers to the north and south of the grandstand to provide a degree of shelter to the terraces. This distinctive lightweight structure is designed to mirror the tree fringe which frames the racecourse.

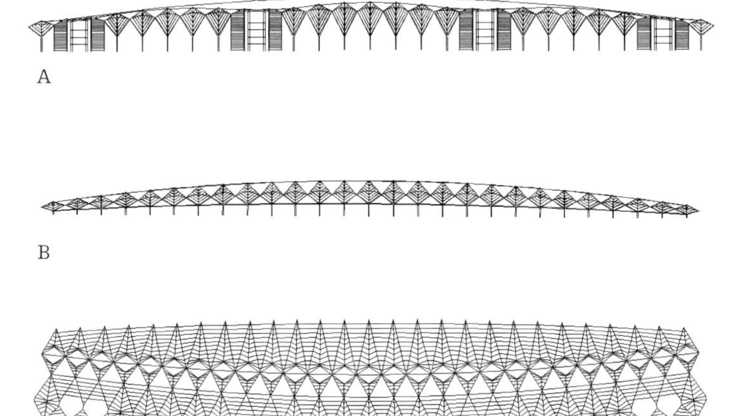

Studies for roof systems (translucent membrane and tubular steel) at Ascot
A) rear
B) front
C) plan

Initially, the aim was to make the monument as high as the 164-metre Great Pyramid of Cheops. In the end, the client agreed to settle for a more manageable 62 metres, but the building was beset by other difficulties. The water table of the site is high and temperatures range from lows of -30 degrees Celsius in winter to +40 degrees Celsius in summer. Speed and the importance of building a light structure made steel construction essential, so very large annual thermal movements are expected. Because of their magnitude, they could not be accommodated by constraining them in a rigid concrete base because internal stresses could be so great that the structure might fail. So the tubular steel frame is tied to itself and mounted on sliding bearings allowing it to expand and contract easily. It is restrained at one point on each face to prevent it slipping off its concrete base; the steel pyramid can flex in- and outwards while being prevented from moving sideways.

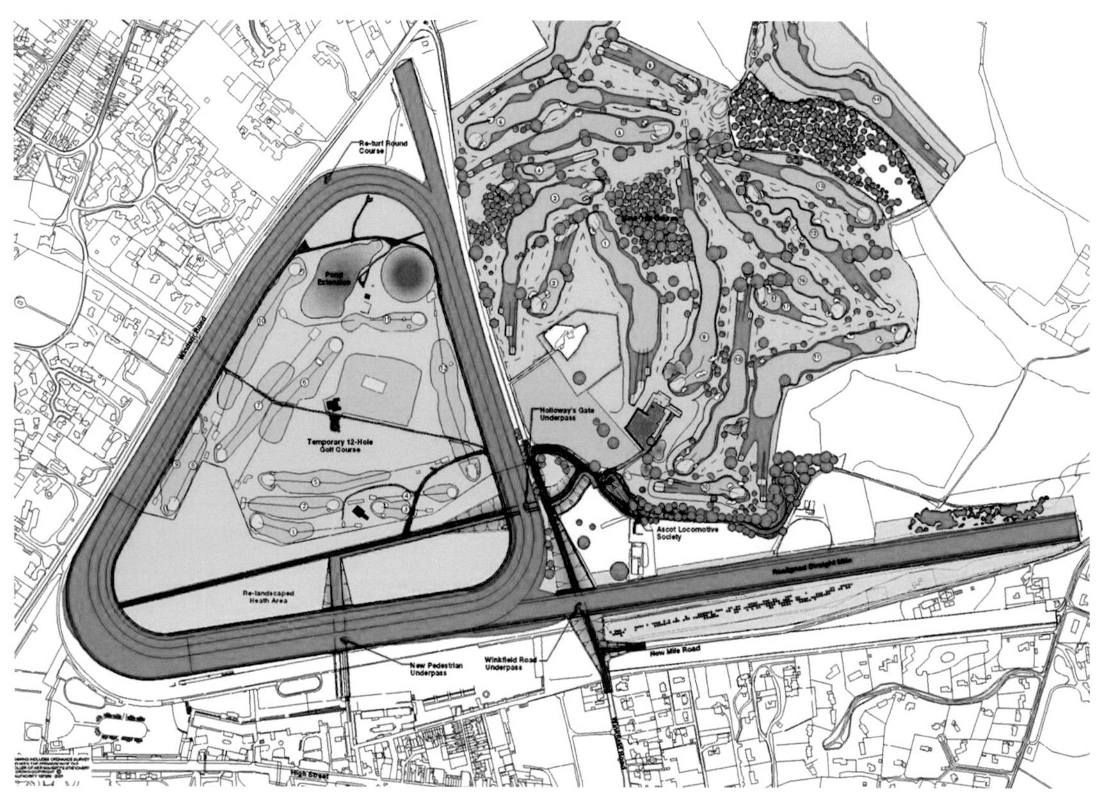

Plan showing re-aligned Royal Mile and new underpass and access to the heath and golf course situated in the centre of the racetrack

In addition to such problems, at the last minute the client requested that an opera house be added to the complex. This meant the creation of a building within the concrete base, with its fly-tower poking up into the pyramid. To avoid going too deeply into the water table, the opera house is created within an artificial mound, so raising the height of the whole monument. The concrete structure of the opera house is independent, both structurally and acoustically, from the steel one. Each had to be built independently, for in Astana concrete can be poured only in summer, while the Turkish-made triangular steel units that form the sides of the pyramid could be bolted together even in the coldest weather. The top of the pyramid (where the international religious delegates meet) is glazed, partly in stained glass. Most of the rest is clad in stone, with prefabricated elements bolted to a watertight inner skin of profiled steel sheet that is itself fixed directly to the tubular steel structure.

Another recent project that had to be completed to a very tight timetable was the new grandstand at Ascot racecourse, designed with architects HOK SVE. Works (including demolition) were started in October 2004 and had to be completed in time for the June Royal Ascot meeting in 2006, 21 months later. Again, prefabrication was used to speed construction, with precast concrete elements forming the main structure. Four shear cores at the back of the stand anchor the entire structure, allowing the rest to be a grid of comparatively thin columns and slabs. The fabric roof is supported on prefabricated lightweight steel trusses, intended to generate a frilly, festal, garden party atmosphere.

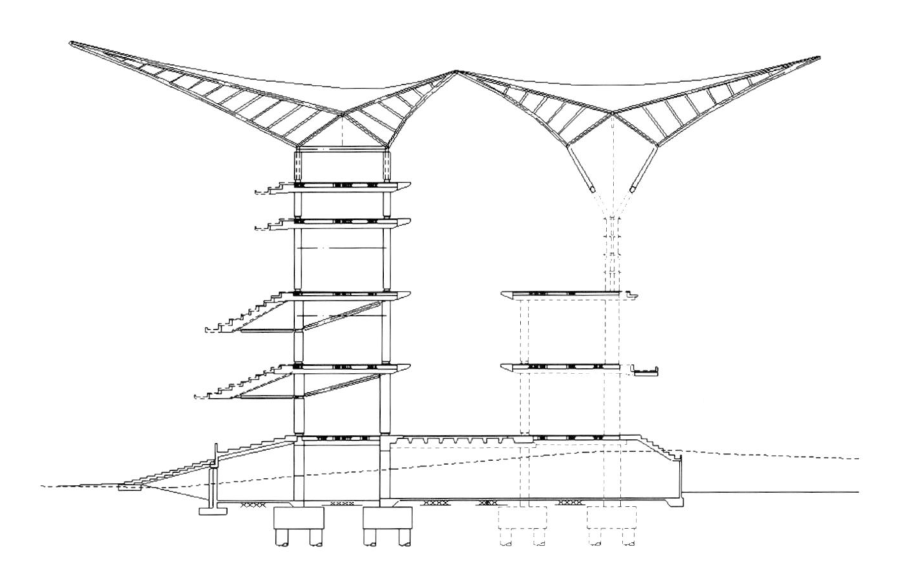

Section through main stands at Ascot showing central atrium and sub-ground servicing level

Perhaps the key criterion in designing buildings from which sports are seen is to ensure that as many spectators as possible have a clear view of the contests and stadium slopes provide compact sightlines to the playing area. Hence, vertical supports must be minimised in both number and section and, where possible, placed out of spectators' field of view. At the new Emirates Stadium for the Arsenal Football Club (again with HOK SVE), just as at Ascot, the primary supports and stabilisers of the terraces and roof are serviced cores, of which there are only eight, some 204 metres apart, in the whole huge amphitheatre. Floor plates are divided into eight zones by movement joints; each zone is attached to the next so that it is stabilised by two cores.

The roof, which shelters all spectators, is the most dramatic part of the structure. Its primary structural elements are lightweight trusses, triangular in section and formed of circular hollow section steel members. A prismatic perimeter truss connects and restrains the radial trusses. Vertical and lateral roof loads are transferred to the concrete structure by tripods that stand on top of each core and by 64 articulated props that transfer loads from the perimeter truss to the rakers of the upper tier of seating. The secondary steel roof structure is covered by two skins that sandwich purlins suspended beneath the exposed primary steelwork. Weather proofing is dealt with by the outer skin; the inner one is formed of aluminium trays that span between the bottom flanges of the purlins. A continuous 10-metre-wide band of polycarbonate sheeting fringes the inner edge of the roof that swoops down towards the pitch, to emphasise the intensity of spectators' experience. This translucent eyebrow is intended to soften shadows that might interfere with clear views of the game on sunny days.

EMIRATES STADIUM

Location: London, UK
Date: 2006
Architect: HOK SVE
Client: Arsenal Football Club
Buro Happold services: Structural engineering, environmental systems design, fire engineering, access design, bridge and civil engineering

The overall form and structural configuration of the Emirates Stadium was derived by striving for a balance between a wide range of factors: high-quality viewing standards; an awe-inspiring atmosphere within the arena; the quality of the pitch microclimate; and the desire for the stadium form to be iconic and instantly recognisable as the home of Arsenal Football Club. Against this, limitations were imposed by the tight site footprint, time and budget constraints and the proximity of the site to existing network rail and London Underground lines.

The resulting configuration is an approximately elliptical-plan form, featuring five main levels of accommodation, and four seating tiers for 60,000 spectators contained within a compact site footprint that also respects the height restriction imposed by local planners. The roof is a naturally dished form that will enhance the microclimate of the pitch and the atmosphere within the arena, and will ensure the stadium is a landmark.

Spectator's view of the playing area beneath translucent inward sloping roof surface

Services are in general divided into eight zones, corresponding to the eight cores. Each has a substation that provides power for catering, and lighting. Services are designed on a decentralised basis to minimise distribution runs and allow use of individual parts of the building for private functions at times when there is no game. Each zone has individual condensing boilers, air cooled chillers and air handling units to cope with the ventilation and cooling loads of densely packed crowds on match days. Plant rooms on the cores at roof level of the upper concourse distribute air, heating and cooling to spaces in the stadium below; air conditioning is restricted to restaurants, while concourses are naturally ventilated. Considerable attention has been paid to the quality of artificial lighting in all the restaurants as well as public spaces.

By using the principal supports as large service ducts, both Ascot and the Emirates Stadium do relate environmental control and structure. But, by being exposed to the open air, stadia are by definition less demanding in terms of environmental control than enclosed spaces. More complex relationships are to be found for instance in the Smithsonian Museum's conversion of the old Patent Office Building in Washington. The building has been refurbished to house the National Portrait Gallery and the National Museum of American Art. As at the British Museum, the central courtyard of a neo-classical monument has been covered by a doubly-glazed roof that allows the resulting enclosed public space to be used for special events and the surrounding wings to be related in a more direct way than by the circulation routes within the classical building.

View of the completed
Emirates Stadium

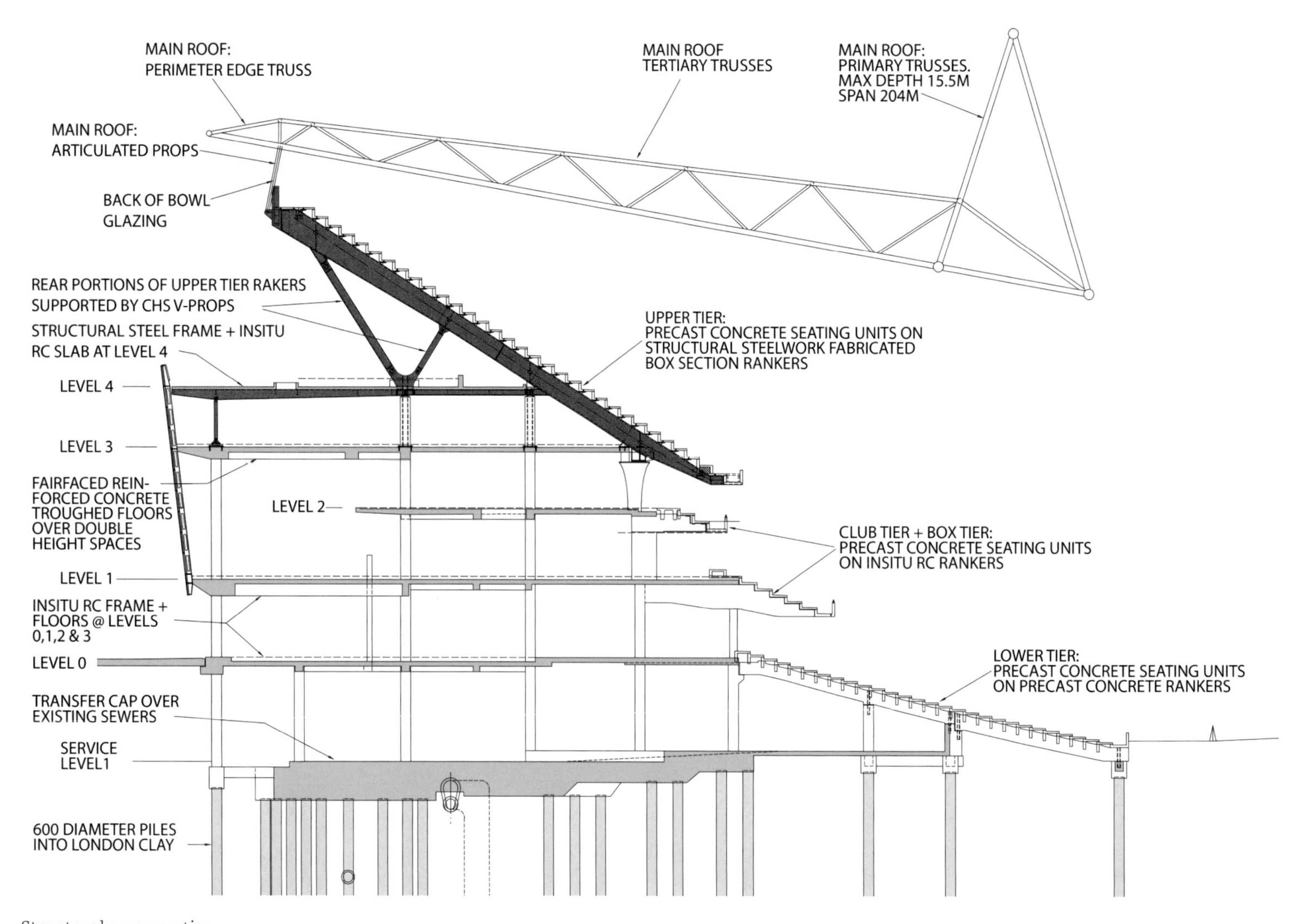

Structural cross-section through main stand showing view lines to the pitch

Aerial view of stadium showing the tight triangular site and new access bridges across railway lines, as well as primary steel trusses to support inward-sloping roof that admits light to the playing area.

1-metre-deep primary triangular truss on tripod and temporary construction trestle

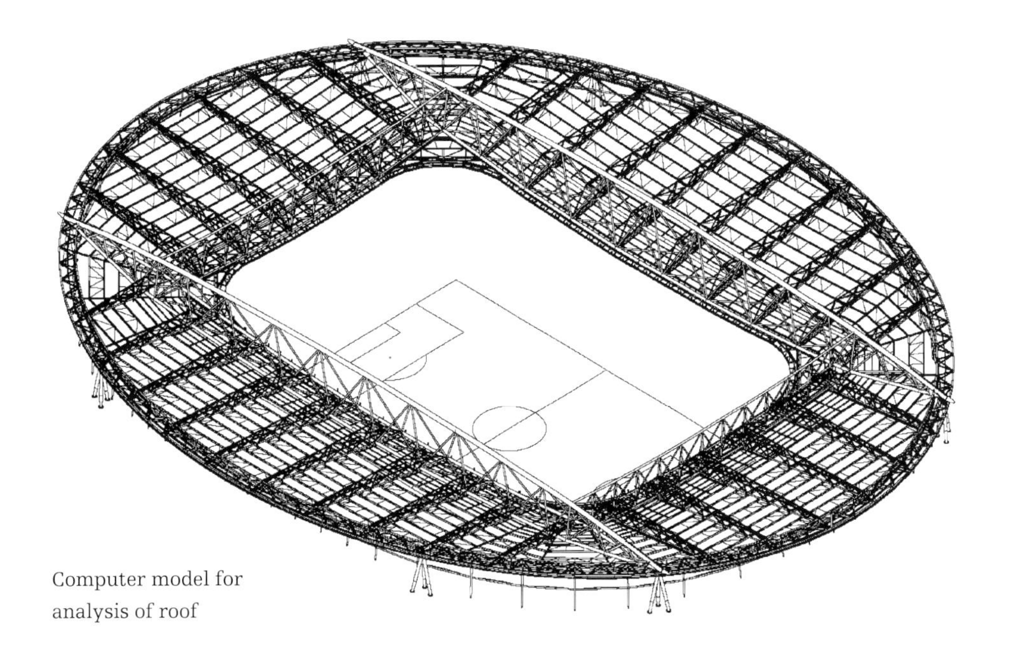

Computer model for analysis of roof

External view

TRINITY BUOY WHARF CONTAINER CITY

Location: Docklands, London, UK
Date: 2001
Architect: Nicholas Lacey & Partners
Client: Urban Space Management
Buro Happold services: Structural engineering

The original Container City project was installed in four days, and completed in five months in 2001. Container City I was originally three storeys high providing twelve work studios across 446 square metres. After high demand a fourth floor was added providing three additional apartments. As well as being very cost-effective Container City I is environmentally friendly with over 80 per cent of the building created from recycled material. To date this alternative method of construction has successfully created youth centres, classrooms, office space, artists studios, live/work space, a nursery and retail space.

Rendering of completed building

SILKEN HOTEL

Location: London, UK
Date: 2008
Architect: Foster + Partners
Client: Grupo Urvasco S.A.
Buro Happold services: Structural engineering, ground engineering, fire engineering, infrastructure (drainage), health and safety, planning supervision

Buro Happold is collaborating with Foster + Partners on this ten-storey hotel in the heart of London, at the east end of the Strand on the site of the Marconi House, the historic building where the first public radio broadcast took place in 1922. The new building will combine restoration of the listed façade of the Marconi House with a new Portland stone building, and will correspond perfectly with the adjacent apartment buildings. The hotel is aiming for a five-star rating and will provide 170 bedrooms and 90 apartments. Restaurants, bars, a rooftop terrace and a central atrium will also be provided. The client is based in Spain and specialises in the development of high-quality hotels.

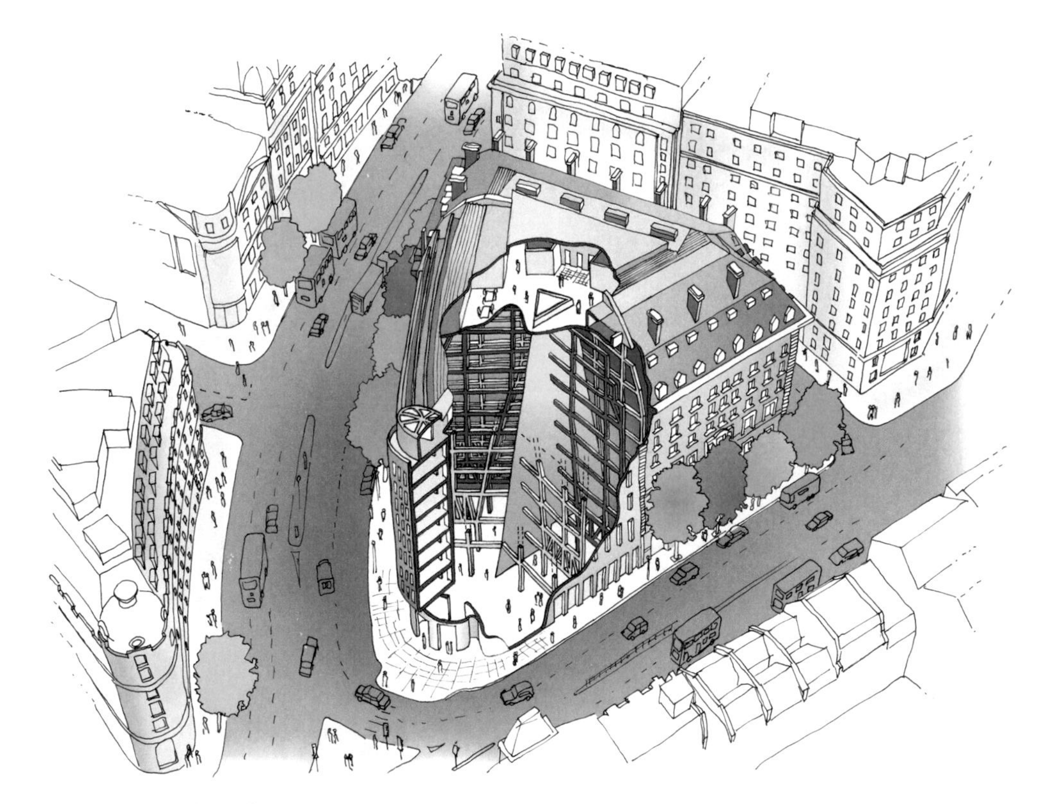

Isometric of atrium of Silken Hotel adjacent to the Aldwych in London

Rendering

View of the Kogod Court at the Smithsonian Patent Office Building, showing the glazed steel grid shell

THE KOGOD COURT AT THE SMITHSONIAN INSTITUTION

Location: Washington, DC, USA
Date: 2007
Architect: Foster + Partners
Client: Smithsonian Institution
Buro Happold services: Structural engineering

The building houses both the National Portrait Gallery and the National Museum of American Art, and underwent a $450 million refurbishment. The structural concept for the new roof has been developed to recognise the importance of environmental control, acoustics, and lighting requirements and respect the fabric of the existing building. Integrating all the technical needs to provide the best environment for the public space led to the development of a sophisticated yet deceptively simple structural idea.

This has an elegant form and allows a highly efficient use of steel, aluminium and glass. The structural design is visually a 'magic carpet' that floats across the courtyard. It comprises three interconnecting vaults that flow into one another via softly curved valleys. These vaults transfer load towards the eight preselected column positions, while the columns connect to one another through the valley zone between vaults.

Working on the Foster + Partners conversion as structural engineers, with Battle McCarthy as environmental consultants, Buro Happold evolved a steel and glass diagrid roof carried on eight large cantilever cylindrical columns which are structurally separate from the surrounding masonry. The diagrid is a two-way lattice of prefabricated steel members that allows the roof form to flow into three gently mounded vaults. These are in compression and direct loads to the column heads; valleys between the mounds connect the heads of the columns laterally. Ties set in the lattice enable it to retain its hovering form and prevent the grillage becoming distorted by temperature changes and imposed stresses of wind and snow. All new or refurbished public buildings in Washington must allow for blast loading. At the old Patent Office Building, the continuous diagrid is predicted to be very resilient in explosive conditions, and because there is a good deal of redundancy in the structure, the roof can remain stable even if part of it is destroyed.

View showing relationship of tubular steel diagonal grid to quadrangular glazing

View of the steel grid shell in construction

Fabric, structure and environmental control are intimately related. The diagrid is made with such precision that glass can be fixed direct to the steel structure without need for secondary glazing bars. Artificial lighting in the cornice of the old masonry building illuminates both the space below and, dramatically, the diagrid above. The tubular roof support columns act as part of the ventilation system, with treated air input just above the courtyard slab and exhaust louvres at high level. Loudspeakers incorporated in the columns allow for music and announcements and potential acoustic manipulation of the space.

Integration is key in Buro Happold's first carbon-neutral building, the Advanced Manufacturing Research Centre (AMRC) designed with architect Bond Bryan for Sheffield University. A steel-braced frame structure reduces the size of internal columns. The stabilising action of the first floor slab is complemented by horizontal trusses in the roof which transfer lateral thrusts to braced bays throughout the building. Translucent and insulating Kalwall cladding, supplemented by ETFE roofing, allows daylight to penetrate deep into the building, so reducing the need for artificial illumination. Careful use of shading prevents excessive solar gain.

External view

Corner detail

As at the Building Research Establishment's offices at Garston, the natural ventilation strategy of the Sheffield building is complemented by exposing the concrete soffits of precast roof planks to act as heat sinks in summer and sources of low-level radiation in cold weather. Energy strategies external to the building include a ground source heat pump that provides low-grade hot water to work the underfloor space heating system. In summer, its machinery can be reversed to circulate chilled water. Mechanical ventilation, where required in the laboratories, is complemented by natural thermal air displacement. The building is to be largely powered by an 86-metre-high wind turbine, capable of generating more than 1 million kilowatt-hours a year. It is key to AMRC's energy strategy: without it, the interlocking climate control systems could not work without constantly drawing on mains electricity. In pursuing this zero carbon goal, much has been learned from post-occupancy observation and measurement of the many green educational buildings that Buro Happold has had the good fortune to work on in the past decade.

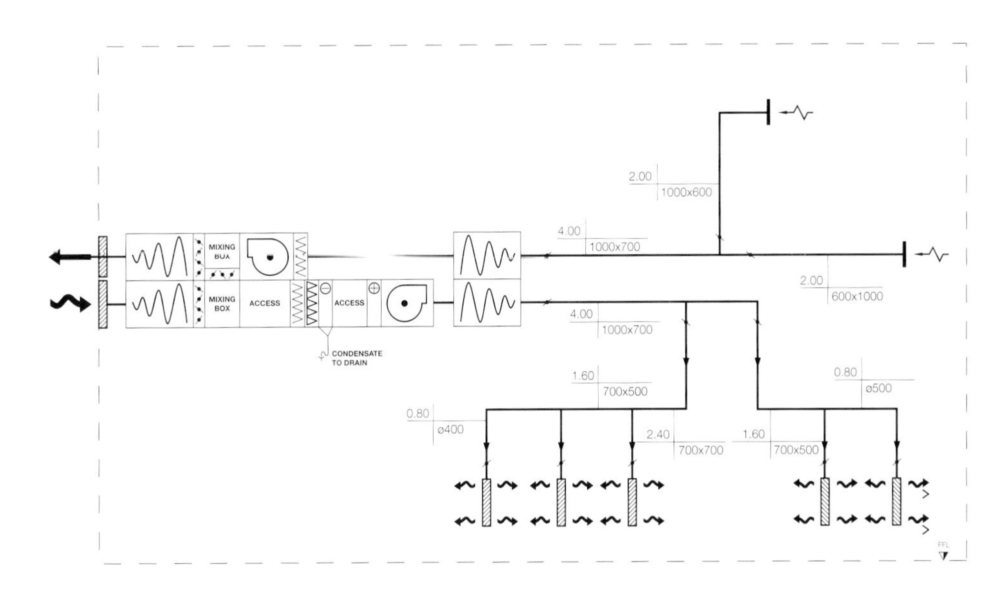

System diagram

ADVANCED MANUFACTURING RESEARCH CENTRE (AMRC)

Location: Sheffield, UK
Date: 2008
Client: University of Sheffield
Architect: Bond Bryan
Buro Happold services: Building services, structural engineering, ground engineering, civil engineering, BREEAM consultation and assessment, acoustics, fire engineering, risk assessment

Located on reclaimed opencast mine land on the outskirts of Sheffield, the Advanced Manufacturing and Research Centre (AMRC) is a world class facility where research, design, manufacture and study can interact effectively. All the AMRC's power is provided by two 250-watt wind turbines that will supply sufficient energy to achieve a zero-carbon building. In addition to this, solutions such as ground source heat pumps (GSHP), natural ventilation and extensive use of daylighting are employed, resulting in a building that reduces both environmental impact and annual running costs.

Early visualisation of AMRC building showing one of the two wind turbines

Another project which uses wind energy is the Wales Institute for Sustainable Education (WISE), a development of the Centre for Alternative Technology (CAT) at Machynlleth in Powys, central Wales. CAT has a well-established reputation for exploring sustainable building techniques and ways of life. Designed with architects Pat Borer and David Lea, WISE is intended to provide facilities for CAT's ever-expanding educational programme. Clearly, it had to incorporate latest thinking in sustainable building construction, energy use, water handling and waste treatment. To reduce carbon dioxide emissions, natural ventilation will be used wherever possible; there will be a biomass-fuelled heat and power system, solar voltaic cells on the roof and extensive use of locally grown timber in structure and cladding.

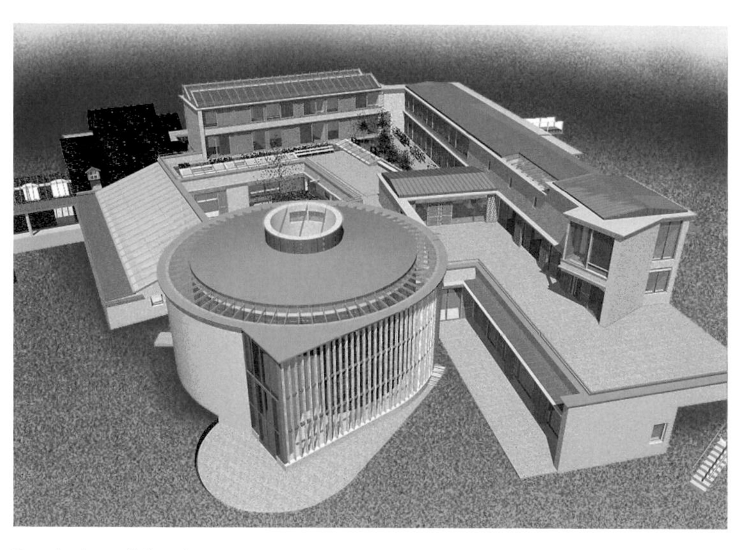

Rendering of the site

WALES INSTITUTE FOR SUSTAINABLE EDUCATION (WISE)

Location: Machynlleth, Powys, Wales, UK
Date: 2007
Architect: Pat Borer and David Lea
Client: Centre for Alternative Technology
Buro Happold services: Structural engineering

The Centre for Alternative Technology (CAT) is launching a new development, the Wales Institute for Sustainable Education WISE. Rammed earth has been chosen for the wall construction of a lecture theatre. This project sets out to be a model of sustainable construction, and will incorporate a number of sustainable technologies and building methods. CAT has previously used rammed earth successfully in the construction of its bookshop, built six years ago. These 3.5-metre walls were straight and stood independently of each other. For its new project, CAT wishes to be more ambitious. The circular wall of the theatre will be 7.5 metres high and 15 metres in diameter, measuring 450 millimetres in width. Utilising some 280 tonnes of soil, this will be one of the largest rammed-earth projects built in the UK.

A circular lecture theatre, the largest element of WISE, is to be made of rammed earth. Its 7.2-metre-high walls are to be made of site subsoil, the clay content of which obviates the need for additional binding material. Walls are packed down in layers between recyclable timber formwork (a traditional technique, but aided by using pneumatic compactors rather than manual rammers). Compared to conventional construction, the technique offers greater thermal mass and its construction is a great deal less energy-intensive. When completed, the theatre will be one of the largest buildings made of earth in recent times.

Section through building featuring main rotunda in rammed earth and timber roof

The shape of the roof has been designed to prevent blocking daylighting from local residents on the north side of the stadium.

LANSDOWNE ROAD STADIUM

Location: Dublin, Ireland
Date: 2009
Architect: HOK SVE and Scott Tallon Walker
Client: Irish Rugby Union and Irish Football Association
Buro Happold services: Ground engineering, planning supervision, structural engineering

The existing stadium will be demolished and the current pitch will be re-aligned to create space for redevelopment of the site which will include a 50,000-seat facility and associated infrastructure. There will be four tiers of seating around three sides of the pitch. The north side of the stadium is close to a residential area and will therefore have only one low-level tier of seating.

All four sides of seating will be protected from the elements by a continuous curvilinear roof; however, the area above the pitch will remain open. An access podium will be created above the railway line allowing spectators to enter the stadium at the second-tier level. In addition, a network of access routes will make it easier and more convenient for spectators to enter and exit the stadium.

Stadium stand arrangement showing concept for principal horse-shoe truss support to roof

Polycarbonate scales form translucent sheath to stadium exterior

External view

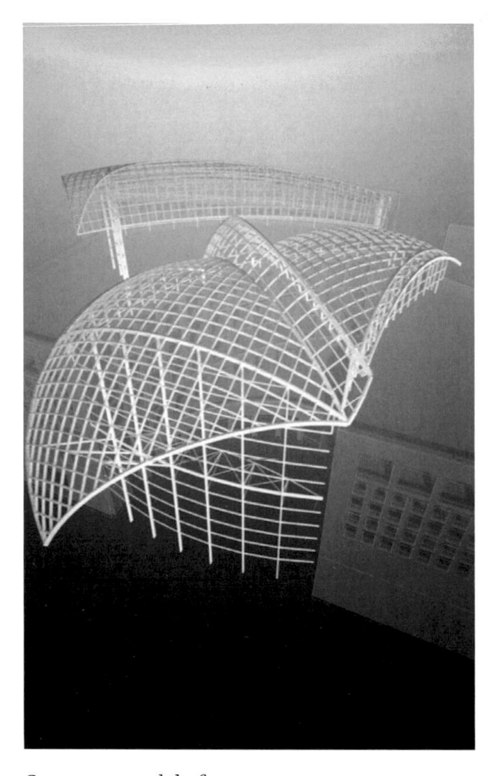

Computer model of structure of steel roof shells

A visualisation of the interior space

Model view

UNITED STATES INSTITUTE OF PEACE HEADQUARTERS

Location: Washington, DC, USA
Date: 2011
Architect: Moshe Safdie and Associates
Client: United States Institute of Peace
Buro Happold services: Structural engineering, building services, computation and simulation analysis

The permanent headquarters building will serve as a national centre for research, education, training, and policy and programme development on issues of international conflict prevention, management and resolution. Moshe Safdie's design concept was approved by the US Commission of Fine Arts in November 2002. In addition to workspace and a research library, the headquarters will include a state-of-the-art conference centre, conference room, and a variety of smaller meeting rooms. The first floor of the building will serve as an education centre.

Elevation

INDEX

ACKNOWLEDGEMENTS

The inspiration for the book originated from Ted Happold and the founding partners Ian Liddell, John Morrison, Peter Buckthorp, Michael Dickson, Rod Macdonald, Terry Ealey and John Reid. These early aspirations have now passed onto a far wider leadership – designers, planners and support staff, too many to mention across many locations in the world – who have discussed and contributed to this publication.

As for the publication itself, Buro Happold's marketing team along with the help of Michael Dickson have kept to the task of communicating energetically with the publishers and designers at Birkhäuser, under the direction of editor Ria Stein. Their endeavours are gratefully acknowledged.

ILLUSTRATION CREDITS

Cover photograph: Buro Happold

Page 10/11: Buro Happold/Paul White
Page 21: Buro Happold/Mandy Reynolds
Page 23: Buro Happold/Mandy Reynolds
Page 24/25: Buro Happold/Paul White
Page 26: Buro Happold/Mandy Reynolds
Page 28: Buro Happold/Hisao Suzuki
Page 29: Buro Happold/Mandy Reynolds
Page 30/31: Buro Happold/Mandy Reynolds
Page 32/33: Buro Happold/Mandy Reynolds
Page 37: Buro Happold/Anton Grassi
Page 38/39: Buro Happold/Jo Poon
Page 40: Buro Happold/Mandy Reynolds
Page 41: Buro Happold/Adam Wilson
Page 42–44: Buro Happold/Alan McAteer
Page 45: Buro Happold/Adam Wilson
Page 47: Buro Happold/Adam Wilson
Page 48: Buro Happold/Alan McAteer
Page 50/51: Buro Happold/Daniel Hopkinson
Page 56: Buro Happold/Mandy Reynolds
Page 59-61: Buro Happold/Mandy Reynolds
Page 63: Buro Happold/Daniel Hopkinson
Page 64/65: Buro Happold/Adam Wilson
Page 66/67: Buro Happold/Mandy Reynolds
Page 69: Buro Happold/Adam Wilson
Page 70 top: Buro Happold/Mandy Reynolds
Page 70 bottom: Hopkins Architects London
Page 72: Buro Happold/Mandy Reynolds
Page 74: Buro Happold/Adam Wilson
Page 76/77: Buro Happold/Adam Wilson
Page 78: Buro Happold/Adam Wilson
Page 80: Buro Happold/Adam Wilson
Page 84: Buro Happold/Adam Wilson
Page 86: Buro Happold/Mandy Reynolds
Page 88: Buro Happold/Alan McAteer
Page 90/91: Buro Happold/Christian Richters
Page 92/93: Buro Happold/Mandy Reynolds
Page 94: Buro Happold/Daniel Hopkinson
Page 95: Buro Happold/Jo Poon
Page 99 top: Buro Happold/Mandy Reynolds
Page 100: National Library Singapore
Page 103: Buro Happold/Michael Moran
Page 104–107: Buro Happold/Werner Huthmacher
Page 109: Buro Happold/Mandy Reynolds
Page 112–113: Buro Happold/Mandy Reynolds
Page 117: Buro Happold/Jo Poon
Page 121: Buro Happold/Adam Wilson
Page 122: Buro Happold/Robert Greshoff
Page 123: Buro Happold/Angus Macdonald
Page 130: Buro Happold/Robert Greshoff
Page 131: Buro Happold/Simon Warren
Page 132: Buro Happold/Sealand Aerial Photography
Page 134: Buro Happold/Robert Greshoff
Page 138: Buro Happold/Paul White

All other illustrations were provided by the Buro Happold image archive.